TRIBOLOGY & DESIGN

WIT*PRESS*

WIT Press publishes leading books in Science and Technology.
Visit our website for the current list of titles.
www.witpress.com

WIT*eLibrary*

Home of the Transactions of the Wessex Institute.
Papers presented at Tribolgy and Design 2010 are archived in the WIT eLibrary in volume 66 of WIT Transactions on Engineering Sciences (ISSN 1743-3533).
The WIT eLibrary provides the international scientific community with immediate and permanent access to individual papers presented at WIT conferences.
Visit the WIT eLibrary at www.witpress.com.

THIRD INTERNATIONAL CONFERENCE ON TRIBOLOGY AND DESIGN

TRIBOLOGY & DESIGN

Conference Chairmen

M. Hadfield
University of Bournemouth, UK

C.A. Brebbia
Wessex Institute of Technology, UK

J. Seabra
University of Porto, Portugal

International Scientific Advisory Committee

Organised by
Wessex Institute of Technology, UK
University of Bournemouth, UK
University of Porto, Portugal

Sponsored by
WIT Transactions on Engineering Sciences

WIT Transactions

Transactions Editor

Carlos Brebbia
Wessex Institute of Technology
Ashurst Lodge, Ashurst
Southampton SO40 7AA, UK
Email: carlos@wessex.ac.uk

Editorial Board

TRIBOLOGY & DESIGN

Editors

M. Hadfield
University of Bournemouth, UK

J. Seabra
University of Porto, Portugal

C.A. Brebbia
Wessex Institute of Technology, UK

WIT*PRESS* **Southampton, Boston**

Editors:

M. Hadfield
University of Bournemouth, UK

C.A. Brebbia
Wessex Institute of Technology, UK

J. Seabra
University of Porto, Portugal

Published by

WIT Press
Ashurst Lodge, Ashurst, Southampton, SO40 7AA, UK
Tel: 44 (0) 238 029 3223; Fax: 44 (0) 238 029 2853
E-Mail: witpress@witpress.com
http://www.witpress.com

For USA, Canada and Mexico

Computational Mechanics Inc
25 Bridge Street, Billerica, MA 01821, USA
Tel: 978 667 5841; Fax: 978 667 7582
E-Mail: infousa@witpress.com
http://www.witpress.com

British Library Cataloguing-in-Publication Data

A Catalogue record for this book is available
from the British Library

ISBN: 978-1-84564-440-6
ISSN: 1746-4471 (print)
ISSN: 1743-3533 (on-line)

The texts of the papers in this volume were set individually by the authors or under their supervision. Only minor corrections to the text may have been carried out by the publisher.

Printed in Great Britain by MPG Books Group, Bodmin and King's Lynn.

Preface

This book contains the papers presented at the Third International Conference in the series "Tribology and Design". It is an exciting development that the Wessex Institute of Technology and the University of Porto have joined the organisation of this conference series that was founded at Bournemouth University. The first two conferences held at Bournemouth concentrated on Tribology related to global environmental issues which provided a forum for academics and professionals to discuss focussed areas of research within a wider context. The remit of the Third International Conference was to consider wider applications of Tribology within the context of engineering design.

Topics in the book include Design Tools, Test Methods, Surface Engineering, Tribology under Extreme Conditions, and Surface Measurements & Lubrication. All these areas will be discussed by delegates from a wide variety of countries. Demand on engineering systems is continually increasing to maintain our quality of life within a world of limited resources. Our understanding of Tribology needs to embrace all aspects of concern including space applications, reliability, energy applications, advanced materials and corrosion. It is essential that academics and professionals have the opportunity to share current ideas concerning Tribology within the widest context.

Mark Hadfield
Algarve, 2010

Contents

Section 3: Surface engineering

Section 4: Tribology under extreme conditions

Section 5: Surface measurements and lubrication

Section 1
Design tools

Tribological modelling of spherical bearings with complex spherical-based geometry and motion

F. C. Wang[1], L. S. Wang[2] & M. L. Sun[2]
[1]*Wuhan University of Science and Technology, China*
[2]*ZWZ (Wafangdian Bearing Group Corporation), China*

Abstract

A general tribological simulation system for the study of spherical bearings with complex spherical-based geometry and under operating conditions was developed in order to simulate friction, lubrication and dynamic contact mechanics, as well as wear on the basis of the regular spherical-grid-data-model (SGDM), where the bearing surface was assumed as being perfectly spherical. The elastic deformation of the complex spherical bearing was evaluated by the equivalent discrete spherical convolution model (EDSC) and the spherical fast Fourier transform technique (SFFT). The representation of either macro- or micro-geometries of spherical bearings in motion and the corresponding simulation of the tribological complexity were performed by the fixed-tracked method (FTM) in the simulation system. Several examples of numerical experiments of transient elastohydrodynamic lubrication, wear prediction and friction, as well as dynamic contact mechanics of spherical bearings with non-sphericity, were investigated in order to validate the tribological simulation system of spherical or aspheric bearings. It was found that the complex tribological problems of the bearings could be resolved by the general methodologies developed in the simulation system.
Keywords: spherical-grid-data-model (SGDM), fixed-tracked method (FTM), equivalent discrete spherical convolution (EDSC), spherical fast Fourier transform (SFFT), spherical bearing, aspheric bearing, non-sphericity, tribological simulation system, elastohydrodynamic lubrication (EHL), mixed lubrication, friction, contact mechanics, wear prediction.

WIT Transactions on Engineering Sciences, Vol 66, © 2010 WIT Press
www.witpress.com, ISSN 1743-3533 (on-line)
doi:10.2495/TD100011

1 Introduction

Spherical bearings have wide applications in industrial engineering and bioengineering, such as spherical bushing, hip joint replacements and spherical joints (Wang *et al.* [1] and Flores *et al.* [2]). The significant progress of tribological modelling of spherical bearings, including friction, wear and lubrication, has been achieved in the past decade ([1] and Wang *et al.* [3]). Most of the simulation works have been carried out by the regular spherical-grid-data-model (SGDM), where the bearing surface is assumed as being perfectly spherical. However, the geometry of a real spherical bearing is complex, which is introduced by the global or local non-sphericity of bearing surfaces in manufacturing, such as machining marks and roughness, and by the modification of bearing geometry due to wear, as well as possible optimisation design, such as texture or non-spherical bearing surface, [1]. The geometry of the real spherical bearing can be described by the geometrical variations relative to the nominal spherical surface. Additionally, the non-sphericity of bearings under the given operating condition can cause complex tribological problems, for example both the dynamic contact mechanics and the computational wear prediction of spherical bearings with a hard-on-hard bearing material combination under the multi-directional transient motion and dynamic loading condition need to consider the geometrical modifications of two bearing surfaces due to wear with time. Therefore, accurate modelling for the study of the complex tribological problems is helpful to understand the tribological mechanism of the real spherical or aspheric bearing.

The aim of this study was to develop a tribological simulation system for the study of the complex tribological problems of the real spherical or aspheric bearing with complex bearing geometry on the basis of the spherical-grid-data-model. Two cases of spherical and non-spherical bearings under the multidirectional transient motion and dynamic loading condition were simulated to validate the simulation system. It was hoped that the general methods with the developed in-house software in the simulation system could be applied to obtain the satisfactory solutions of the complex tribological problems under the given operating conditions, including friction, lubrication and mixed lubrication, dynamic contact mechanics and wear.

2 Bearing geometry and representation

The geometry of a real spherical bearing or spherical bushing can be described as shown in figs. 1(a) and 1(b). The applied motion and loading to the bearing are described in the appropriate spherical coordinate system, as shown in figs. 2 and 3. When deformation is considered, the complex geometry of a real spherical or aspheric bearing can be described by the geometrical components with the different scales as follows:

$$r_i = r_{i0}(\varphi,\theta) + \Delta r_i(\varphi,\theta) + \Theta_i(\varphi,\theta) + \delta_i(\varphi,\theta) \tag{1}$$

where the subscript r_0 represents the radius of nominal spherical bearing surface. The non-spherical variations of an aspheric bearing can be represented in the spherical coordinates by the deviations, Δr_i, relative to the nominal spherical surface, as shown in figs. 4(a) and 4(b), while Θ_i is the micro-geometrical change such as roughness or machining marks, relative to the overall smooth bearing geometry such as ellipsoidal bearing surface. The elastic deformation of bearing can be evaluated by the equivalent discrete spherical convolution model (EDSC) combined with the corresponding spherical fast Fourier transform (SFFT), [1] and Wang *et al.* [4].

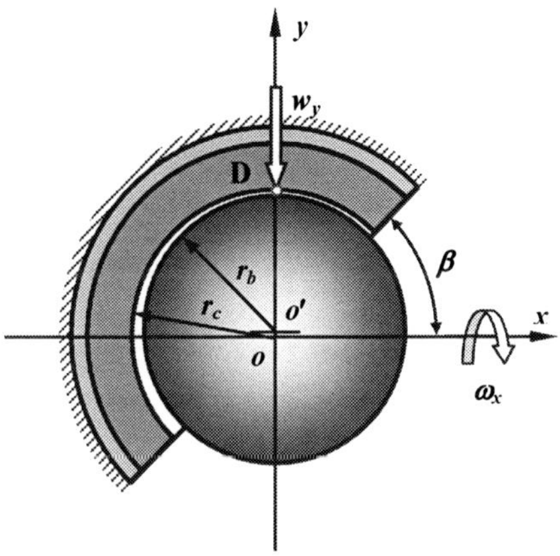

(a)

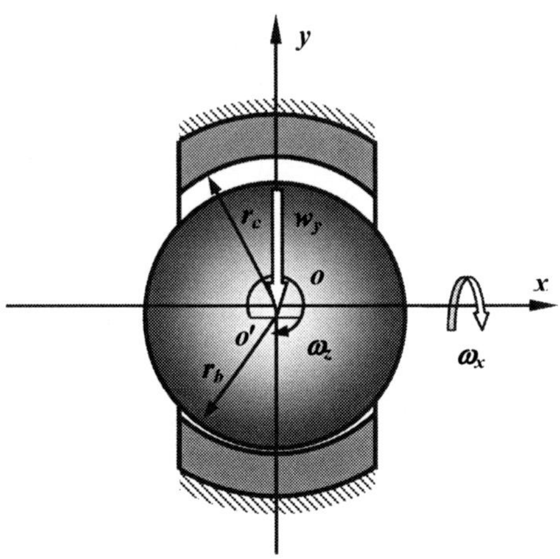

(b)

Figure 1: Geometry model of spherical bearing (a) and spherical bushing (b).

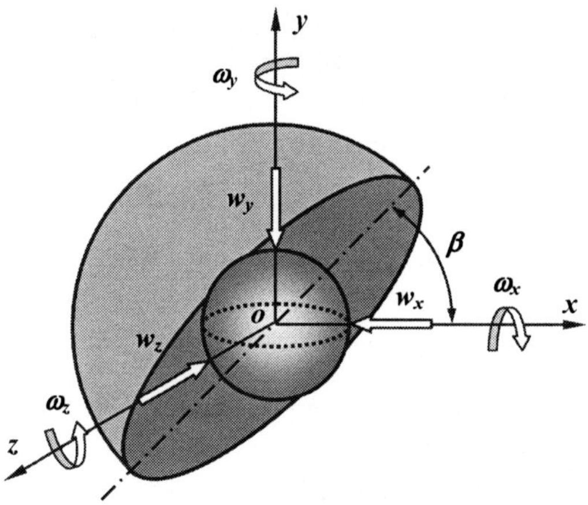

Figure 2: Spherical bearing under loading and motion.

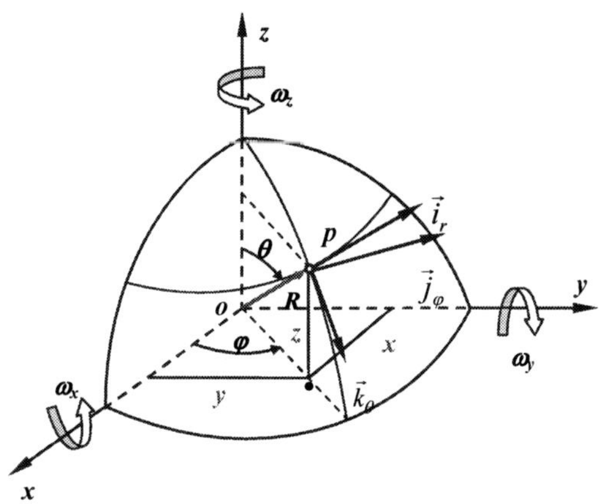

Figure 3: Spherical coordinate system.

Real bearing surfaces need to be accurately described by the discrete spherical mesh-grid space in spherical coordinate system during simulations, for example the geometry representation of complex or moving bearing surfaces is required for the tribological modelling such as the transient lubrication of aspheric bearing and the computational wear modelling of spherical bearing with the geometry modification. For such problems, the fixed-tracked method (FTM) developed recently can be employed to carry out the data communication in the geometry representation at each instant during numerical simulations. Therefore, the two sets of mesh-grids for the numerical simulation and the description of the moving interface of bearing are designed ([1] and Wang *et al.* [5, 6]). The coordinate location of the corresponding mesh grids is determined by

$$[\mathbf{r}] = [\mathbf{A}]^{-1}[\mathbf{r}'] \tag{2}$$

where $[\mathbf{r}]$ and $[\mathbf{r}']$ are the location vectors at a point on the bearing surface in the space and body coordinate systems, and the inverse Euler angular transformation in eqn. (2) is given as follows:

$$A^{-1} = \begin{pmatrix} \cos\beta\cos\gamma & -\sin\beta & \cos\beta\sin\gamma \\ \cos\alpha\sin\beta\cos\gamma & \cos\alpha\cos\beta & \sin\alpha\cos\beta \\ \sin\alpha\sin\beta\cos\gamma & \sin\alpha\cos\beta & \sin\alpha\sin\beta\sin\gamma+\cos\alpha\cos\gamma \end{pmatrix} \tag{3}$$

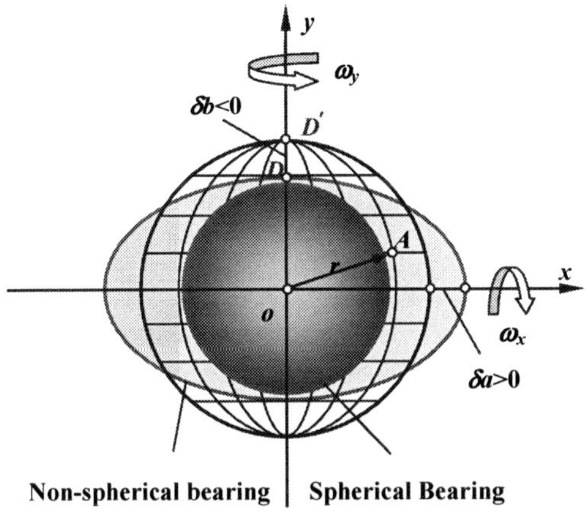

(a)

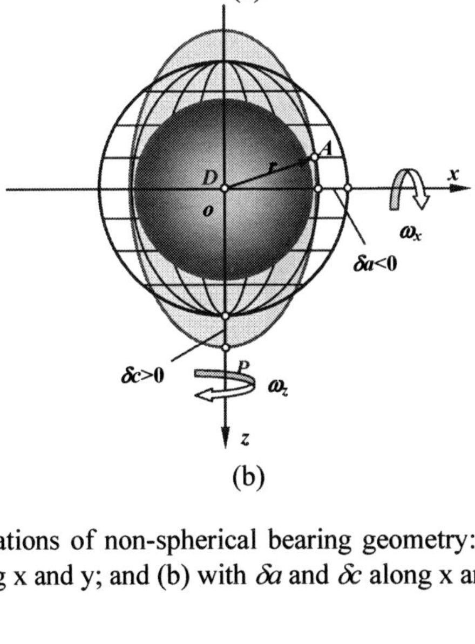

(b)

Figure 4: Variations of non-spherical bearing geometry: (a) with δa and δb along x and y; and (b) with δa and δc along x and z.

where (α, β, γ) are three Euler angles corresponding respectively to the three Euler angular velocities (ω_x, ω_y, ω_z), as shown in fig. 2, [1, 4]. Additionally, the film thickness or gap between the bearing surfaces with the complex geometry such as non-sphericity or roughness can be described as follows [1]:

$$\begin{aligned} h(\varphi,\theta,t) = (r_c - r_b) - e_x(t)\sin\theta\cos\varphi - e_y(t)\sin\theta\sin\varphi - e_z(t)\cos\theta \\ + \Theta_{cb}(\varphi,\theta) + \delta_{cb}(\varphi,\theta,t) \end{aligned} \tag{4}$$

where r_b and r_c are the radial distance from a point on the inner or outer bearing surface of the real spherical or aspheric bearing to the centre of the nominal spherical bearing, respectively. The variables of (e_x, e_y, e_z) are the position of the centre of the inner bearing surface with reference to the centre of the outer bearing surface. The variables of Θ_{cb} and δ_{cb} are the combined micro-geometrical variation and deformation of two bearing surfaces, respectively, Wang *et al.* [7].

3 Governing equation

3.1 Lubrication

The lubrication problem of spherical or aspheric bearing with complex geometry includes elastohydrodynamic lubrication and mixed lubrication under the transient motion and dynamic loading condition. The lubrication film pressure can be evaluated as follows:

$$\begin{aligned} &\sin\theta\frac{\partial}{\partial\theta}\left(h^3\sin\theta\frac{\partial p}{\partial\theta}\right) + \frac{\partial}{\partial\varphi}\left(h^3\frac{\partial p}{\partial\varphi}\right) \\ &= 6\mu r_{c0}^2\sin\theta\left[\begin{aligned} &(-\omega_x\sin\varphi + \omega_y\cos\varphi)\frac{\partial h}{\partial\theta} \\ &+(-\omega_x\cos\varphi\cos\theta - \omega_y\sin\varphi\cos\theta + \omega_z\sin\theta)\frac{\partial h}{\partial\varphi}\end{aligned}\right] \\ &+ 12\mu r_{c0}^2\sin^2\theta\frac{\partial h}{\partial t} \end{aligned} \tag{5}$$

where r_{c0} is the radius of the nominal spherical bearing of the bearing. For the mixed lubrication regime, the asperity contact pressure needs to be evaluated by direct contact mechanic model or other methods as described previously [1, 3, 7].

3.2 Friction and frictional torque

Shear stress of fluid film of spherical or aspheric bearing dependent on velocity change of fluid film and lubricant viscosity can be expressed as follows:

$$[\tau]_f = \begin{Bmatrix} \tau_{\theta\varphi} \\ \tau_{\theta r} \\ \tau_{\varphi r} \end{Bmatrix} = \begin{bmatrix} 0 \\ -\mu \dfrac{U_\varphi}{h} + \left(\varsigma - \dfrac{h}{2}\right) \dfrac{1}{r_0 \sin\theta} \dfrac{\partial p}{\partial \varphi} \\ -\mu \dfrac{U_\theta}{h} + \left(\varsigma - \dfrac{h}{2}\right) \dfrac{1}{r_0} \dfrac{\partial p}{\partial \theta} \end{bmatrix} \tag{6}$$

where U_φ and U_θ, represent the corresponding velocity components at a point on the bearing surface.

The corresponding frictional torque caused by both the fluid film shear stress and the asperity contact can be calculated by

$$\begin{Bmatrix} T_x \\ T_y \\ T_z \end{Bmatrix} = \iint_{S_f} \begin{bmatrix} (\tau_{\theta r} \sin\varphi + \tau_{\varphi r} \cos\varphi \cos\theta) \\ (\tau_{\theta r} \cos\varphi + \tau_{\varphi r} \sin\varphi \cos\theta) \\ (\tau_{\varphi r} \sin\varphi) \end{bmatrix} r_0^3 \, d\theta d\varphi + \iint_{S_b} p f_c \begin{bmatrix} r_{ox}\kappa_x \\ r_{oy}\kappa_y \\ r_{oz}\kappa_z \end{bmatrix} r_0^2 \, d\theta d\varphi \tag{7}$$

where κ_x, κ_y and κ_z are equal to 1 or 0, depending on whether there is relative motion, f_c is the friction coefficient and r_{ox}, r_{oy} and r_{oz} are the corresponding arm lengths of friction stress components to the corresponding coordinate axes [3].

3.3 Contact mechanics

The general description for contact problem of spherical or aspheric bearing with complex geometry can be made by the restriction relationship of the gap between the two surfaces brought into contact and the corresponding contact pressure in the contact area Ω_c as follows:

$$\begin{cases} g(\varphi,\theta) = 0 & (\varphi,\theta) \in \Omega_c \\ p(\varphi,\theta) > 0 & (\varphi,\theta) \in \Omega_c \\ g(\varphi,\theta) > 0 & (\varphi,\theta) \notin \Omega_c \\ p(\varphi,\theta) = 0 & (\varphi,\theta) \notin \Omega_c \end{cases} \tag{8}$$

where p represents contact pressure, and the gap of g comprises the original unreformed gap g_0 without contact pressure and the deformation caused by contact pressure on the two surfaces ([7] and Wang *et al.* [8]). As an example, the contact mechanics was numerically predicted by the finite element method in this study, as shown in fig. 5, in order to validate the general methodology of data communication by the fixed-tracked method for the dynamic contact mechanics as well as the computational wear modelling.

3.4 Computational wear prediction

Once the contact pressure is obtained, the computational wear modelling can be carried out by the fixed-tracked method during the multi-directionally transient motion and dynamic loading condition. The wear depth evaluated by Archard's wear law, experienced an update stage with n_t cycles, can be written as

$$\Delta r_i(t) = \sum_{t=1}^{n_t} k p_t s_t \tag{9}$$

During the operating condition, the bearing geometry is modified with time due to wear. The sliding distance in spherical coordinate system is shown in fig. 6. It should be pointed out that the modification of bearing geometry occurs on the two bearing surfaces of the bearing with a hard-on-hard material combination such as the metal-on-metal bearing [7]. The modifications of the two bearing surfaces at each instant during the operating cycle are expressed as

$$r_i^k(t) = r_i^{k-1}(t) \pm \Delta r_i(t) \tag{10}$$

where + or – sign presents the direction of modification of bearing geometry during the computational wear simulation.

Figure 5: Finite element model for dynamic contact mechanics.

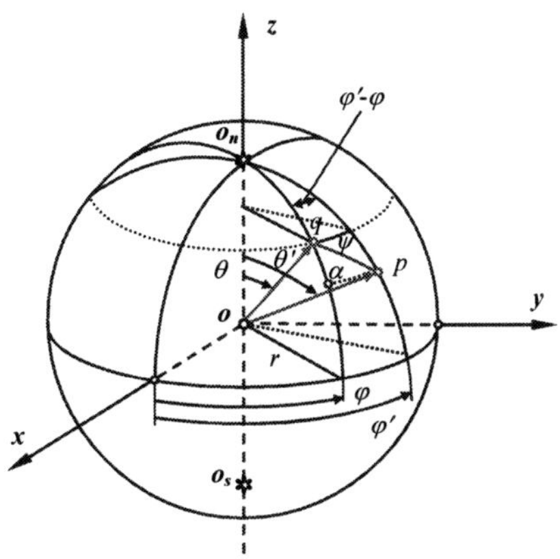

Figure 6: Spherical distance on bearing surface.

4 Numerical experiments

The two cases of complex spherical bearings were carried out for the validation of the tribological simulation system, one for the lubrication and friction of an ellipsoidal bearing with the non-sphericity relative to the nominal spherical bearing in the x-, the y- and z-directions and the other for the computational wear prediction of a spherical bearing. The one-dimensional time-dependent motion and dynamic loading condition was employed for the study of the transient elastohydrodynamic lubrication of the non-spherical bearing. The details of the basic parameters of bearing geometry and the mechanical properties of bearing components had been reported elsewhere [1].

Firstly, the transient elastohydrodynamic lubrication of the ellipsoidal bearing geometry with the non-sphericity of (δa, δb, δc) = (+6, -6, +6) μm was evaluated by running the in-house software in the tribological simulation system with incorporating the general methodologies and the corresponding numerical techniques [1]. The variation of the predicted minimum film thickness with time in three operating cycles is shown in fig. 7. It was seen that the variation became convergent from the second cycles.

Additionally, the prediction of frictional torque caused by shear stress of fluid film of the non-spherical bearing with the ellipsoidal geometry was made when the lubrication result has been achieved under the same operating condition as the above-mentioned, as shown in fig. 8. It was found that the frictional torque at the instants with zero value of angular velocity became zero. This reason follows that friction only occurs in fluid film with relative velocity or velocity grade [3]. The friction stress in the mixed lubrication regime was caused by the asperity contact, which will be studied in future.

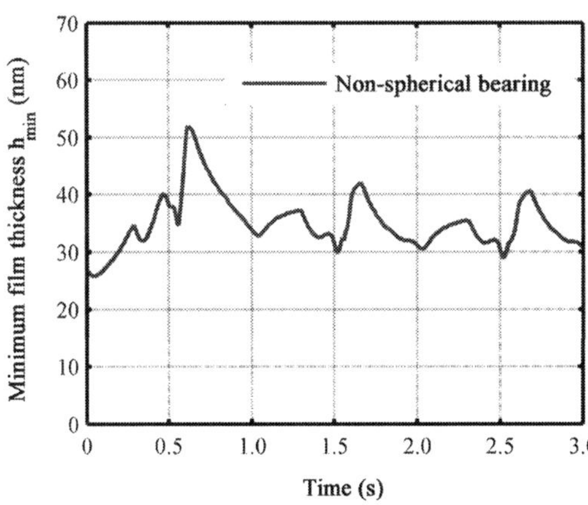

Figure 7: Minimum film thickness.

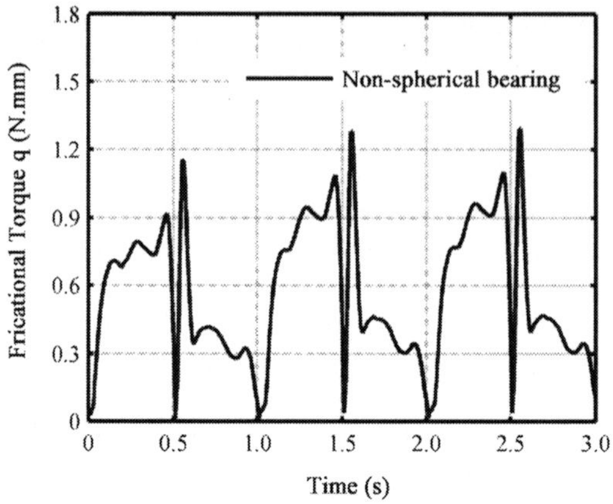

Figure 8: Frictional torque.

Furthermore, the computational wear prediction of metal-on-metal spherical bearing under the three-dimensional time-dependent motion and dynamic loading condition was carried out. The dynamic contact mechanics was evaluated by the tribological simulation system with the finite element model, as shown in fig. 5. The contact pressure distributions at several instant in one operating cycle were presented, as shown in figs. 9(a)-9(d). At the same time, the computational wear prediction up to 10 year (1 million cycles) was carried out, as shown in fig. 10. It should be pointed out that the micro contact mechanics model had to be used for wear prediction of spherical bearing with the micro-geometrical variations such as roughness because of limitation of element number of the finite element method [1, 7, 8].

5 Conclusions

A general tribological simulation system for the study of complex problems of spherical or aspheric bearing with the complex geometry and under the complex time-dependent motion and dynamic loading condition has been developed, in which the general methodologies and numerical techniques were incorporated on the basis of the regular spherical-grid-data-model. Two cases of an ellipsoidal bearing for the study of lubrication and friction and another spherical bearing for the study of dynamic contact mechanics and computational wear prediction under the given operating conditions were investigated. It was seen that the tribological simulation system could be employed to investigate the similar complex tribological problems of spherical or aspheric bearing with complex bearing geometry and under the complex operating condition.

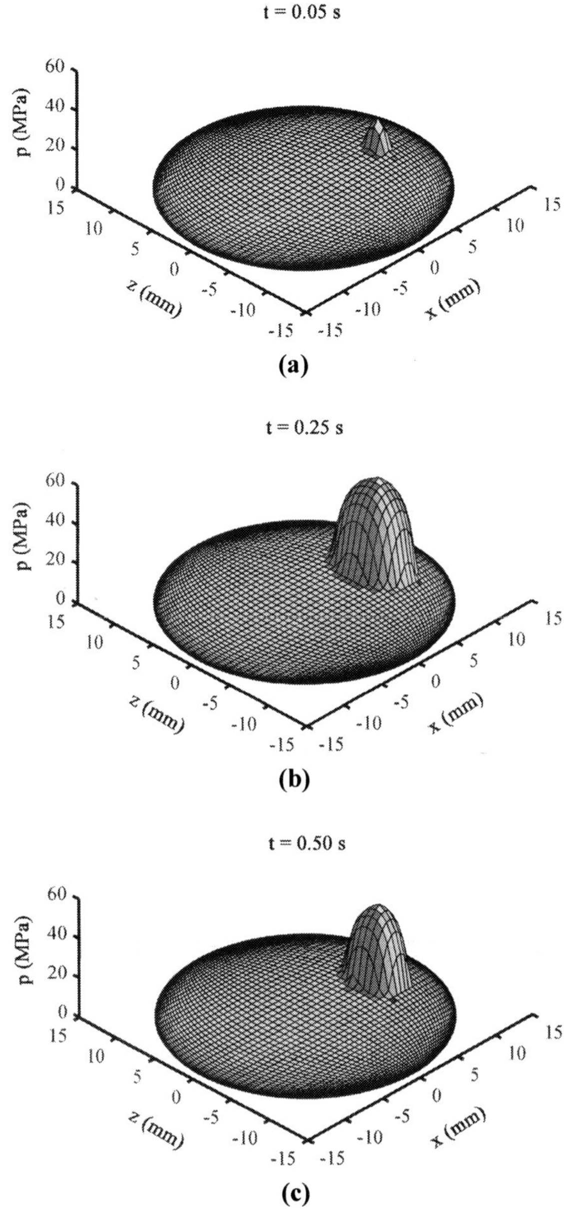

Figure 9: Dynamic contact pressures.

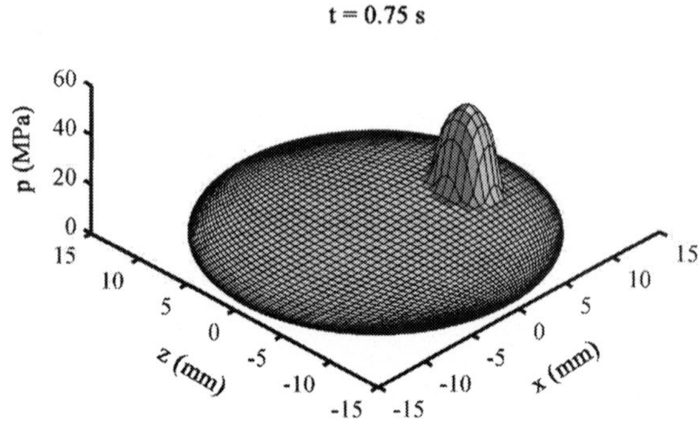

(d)

Figure 9: Continued.

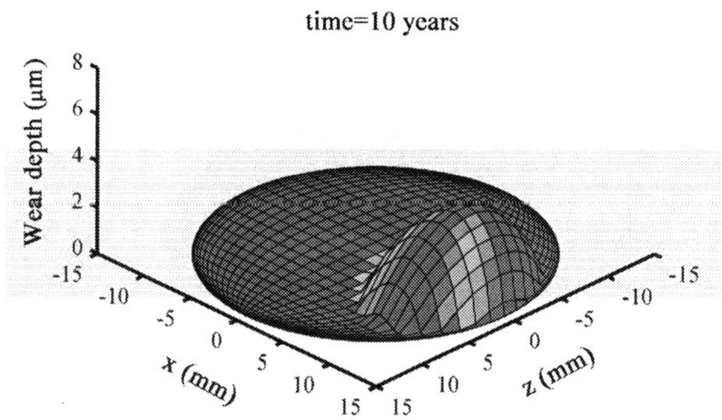

Figure 10: Computational wear prediction.

Acknowledgement

This work was supported by the National Science Foundation of China (Grant No.10972165).

References

[1] Wang, F. C., Zhao, S.X., Félix Quiñonez, A., Hu, X., Mei, X.S. & Jin, Z., Nonsphericity of bearing geometry and lubrication in hip joint implants, *ASME Journal of Tribology*, **131**(3), p031201, 2009.

[2] Flores, P., Ambrósio, J., Claro, J.C.P. & Lankarani, H.M., Dynamics of multibody systems with spherical clearance joints, *ASME Journal of Tribology*, **1**(7), pp. 240-247, 2006.

[3] Wang, F.C, Brockett, C., Williams, S., Udofia, I.T., Fisher, J. & Jin, Z., Lubrication and friction prediction in metal-on-metal hip Joint implants, *IOP Physics in Medicine and Biology*, **53**, pp.1277-1293, 2008.
[4] Wang, F. C. & Jin, Z., Transient elastohydrodynamic lubrication of hip joint implants, *ASME Journal of Tribology*, **130** (1), p011007, 2008.
[5] Wang, F. C., Zhao, S.X., Gao, Q.J., Fan, Q., Zeng, L.C., Félix Quiñonez, A. & Keogh, P.S., Non-spherical bearing geometry and elastohydrodynamic lubrication of hip joint replacements under transient walking conditions, *World Tribology Congress IV*, Kyoto, Japan, 2009.
[6] Wang, F. C., Zhao, S.X., Xu, H., Chen, W., Mei, X.S. & Brown, T.D., Dynamic contact mechanics and wear modelling of hip joint replacements with hard-on-hard material combination under three-dimensional Loading and transient motion, *World Tribology Congress IV*, Kyoto, Japan, 2009.
[7] Wang, F. C., Udofia, I.T. & Jin, Z., An integrated experimental and theoretical contact mechanics study of UHMWPE hip implants tested in a hip simulator, *ASME Proc. World Tribology Congress III*, Washington, pp311-312, 2005.
[8] Wang, F. C. & Jin. Z., Lubrication modelling of artificial hip joints, IUTAM symposium on elastohydrodynamic and micro- elastohydrodynamic 134: book series: *Solid Mechanics and its Applications*, Eds by Snidle RW and Evans HP, Springer, pp385-396, 2006.

General static load capacity in four contact point slewing bearings

J. Aguirrebeitia, R. Avilés, I. F. de Bustos & M. Abasolo
Department of Mechanical Engineering,
University of the Basque Country, Spain

Abstract

This work tries to represent the general static load-carrying capacity of four-contact-point slewing bearings under general loading. This representation is based on a generalization of Sjoväll and Rumbarger's equations and provides an acceptance surface in the load space. This acceptance surface provides the key set of acceptance curves for the design and selection of bearings.
Keywords: four contact point slewing bearing, acceptance surface, bearing selection, static load capacity.

1 Introduction

Slewing bearings are large-sized bearings with many applications, such as in wind turbine generators, tower cranes vertical lathe tables... The loads acting on these bearings usually contain axial and radial forces, as well as tilting moments (see Figure 1). In the most unfavourable load case, the radial force is perpendicular to the resultant of the tilting moments. Several bearing manufacturers provide acceptance curves that allow one to determine whether or not a bearing is acceptable for a given equivalent load, calculated as a combination of the axial and radial loads. By means of a moment–axial-force diagram, this equivalent load allows a designer to obtain the maximum allowable tilting moment that the bearing can bear. This is illustrated in Figure 2. There are some variations in the form and limits of the diagram shown in Figure 2. depending on the manufacturers having experimented with or assessed the bearings themselves, or having simply copied other manufacturers' data. Anyhow, there is always a certain ambiguity and a lack of a clear criterion in the

WIT Transactions on Engineering Sciences, Vol 66, © 2010 WIT Press
www.witpress.com, ISSN 1743-3533 (on-line)
doi:10.2495/TD100021

definition of the equivalent load. The objective of this work is the development of a procedure which defines a surface by the limiting values of the loads F_A, F_R and M for four-contact-point slewing bearings. This representation can be used directly to determine whether or not a given load combination is acceptable.

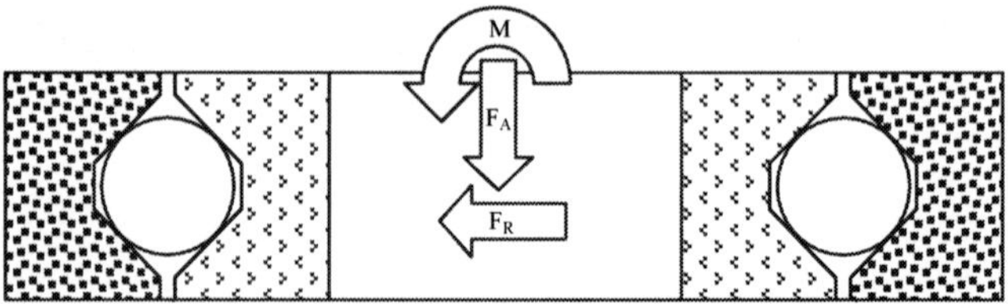

Figure 1: Loads acting on a slewing bearing.

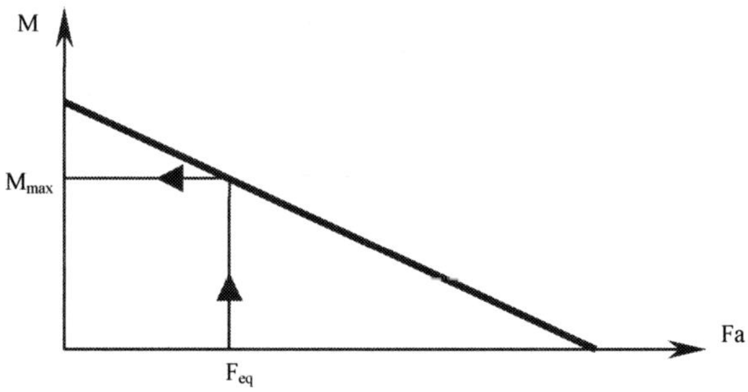

Figure 2: Moment–axial-force diagram for a four-contact-point slewing bearing.

There have been some previous publications where concepts relevant to the assessment of the static load-carrying capacity of four-contact-point slewing bearings have been examined [1-5]. All of the above papers propose a generalization of the equations obtained by Jones [6], in which the load distribution is worked out from the known external loads, taking account of the variation in contact angle with the loading conditions.

This work has a different focus, consisting in directly calculating the load combinations that result in static failure (as defined in the ISO standard [7]) of the most loaded element. This allows one to obtain a three-dimensional acceptance condition in the form of a surface inequation. The designer can use this acceptance surface as a straightforward way to select a bearing appropriately. This approach is based on the calculations of Sjovåll [8] for combinations of axial and radial loads and of Rumbarger [9] for combinations of axial and moment loads. These calculations assume zero clearance in the contact, and rigid rings. These assumptions are also made in the current paper. The axial load-carrying capacity is used to normalize the results and can be obtained from standards [7].

2 The three-dimensional acceptance condition

As in [8] and [9], the first step consists on arranging a geometrical interference model, and then, taking the interference field as deformation field, equations that reflect the equilibrium of the forces and moments are worked out. Finally, the equilibrium equations are rewritten to provide an acceptance inequation. It has been assumed that only the inner ring can be displaced, while the outer one and the balls remain fixed. The deformation field can be expressed as follows, measured as displacements perpendicular to raceways 1 and 2:

$$\begin{aligned} \delta_\psi^1 &= \delta_a \sin\alpha + \delta_r \cos\alpha\cos\psi + \frac{\theta d}{2}\sin\alpha\cos\psi \\ \delta_\psi^2 &= -\delta_a \sin\alpha + \delta_r \cos\alpha\cos\psi - \frac{\theta d}{2}\sin\alpha\cos\psi \end{aligned} \tag{1}$$

where ψ is the angular location over the rings, α the contact angle, d is the contact diameter and δ_a, δ_r, θ are the axial radial and angular displacements of the inner ring with respect the outer one. Superscript 1 refers to the upper raceway and superscript 2 refers to the lower one.

We define auxiliary variables A, R and M as follows:

$$\begin{aligned} A &= \delta_a \sin\alpha \\ R &= \delta_r \cos\alpha \\ M &= \frac{\theta d}{2}\sin\alpha \end{aligned} \tag{2}$$

Using equation (2), equation (1) can be written as:

$$\begin{aligned} \delta_\psi^1 &= A + (R+M)\cos\psi \\ \delta_\psi^2 &= -A + (R-M)\cos\psi \end{aligned} \tag{3}$$

The maximum and minimum values of the deformation fields occur when $\psi = 0$ and $\psi = \pi$, so that

$$\begin{aligned} \delta_0^1 &= A + R + M \\ \delta_\pi^1 &= A - R - M \\ \delta_0^2 &= -A + R - M \\ \delta_\pi^2 &= -A - R + M \end{aligned} \tag{4}$$

If we rewrite equations (12) in terms of the values of the deformation for $\psi = 0$ and $\psi = \pi$:

$$\begin{aligned}\delta_\psi^1 &= \frac{1}{2}\left[\left(\delta_0^1 + \delta_\pi^1\right) + \left(\delta_0^1 - \delta_\pi^1\right)\cos\psi\right] \\ \delta_\psi^2 &= \frac{1}{2}\left[\left(\delta_0^2 + \delta_\pi^2\right) + \left(\delta_0^2 - \delta_\pi^2\right)\cos\psi\right]\end{aligned} \tag{5}$$

According with expressions developed in [8] and [9], and after some mathematical adaptations, action equilibrium equations can be written as shown below, where contributions of raceways 1 and 2 are considered.

$$\begin{aligned}\frac{F_A}{Q_{MAX} Z \sin\alpha} &= F_Q^1 J_I\left(\delta_0^1, \delta_\pi^1\right) - F_Q^2 J_I\left(\delta_0^2, \delta_\pi^2\right) \\ \frac{F_R}{Q_{MAX} Z \cos\alpha} &= F_Q^1 J_{II}\left(\delta_0^1, \delta_\pi^1\right) + F_Q^2 J_{II}\left(\delta_0^2, \delta_\pi^2\right) \\ \frac{M/d}{Q_{MAX} Z \sin\alpha} &= F_Q^1 \frac{1}{2} J_{II}\left(\delta_0^1, \delta_\pi^1\right) - F_Q^2 \frac{1}{2} J_{II}\left(\delta_0^2, \delta_\pi^2\right)\end{aligned} \tag{6}$$

where

$$\begin{aligned}J_I\left(\delta_0, \delta_\pi\right) &= \frac{1}{2\pi} \int_{\delta+} \left(\frac{\delta_\psi}{\delta_{MAX}}\right)^{\frac{3}{2}} d\psi \\ J_{II}\left(\delta_0, \delta_\pi\right) &= \frac{1}{2\pi} \int_{\delta+} \left(\frac{\delta_\psi}{\delta_{MAX}}\right)^{\frac{3}{2}} \cos\psi \, d\psi\end{aligned} \tag{7}$$

and

$$\begin{aligned}&\text{if} \left\{\begin{matrix}\delta_0^1 = \max\left\{\delta_0^1, \delta_\pi^1, \delta_0^2, \delta_\pi^2\right\} \\ \text{or} \\ \delta_\pi^1 = \max\left\{\delta_0^1, \delta_\pi^1, \delta_0^2, \delta_\pi^2\right\}\end{matrix}\right\} \text{then} \left\{\begin{matrix} F_Q^1 = 1 \\ F_Q^2 = \left(\dfrac{\max\left\{\delta_0^2, \delta_\pi^2\right\}}{\max\left\{\delta_0^1, \delta_\pi^1\right\}}\right)^{\frac{3}{2}}\end{matrix}\right\} \\ &\text{if} \left\{\begin{matrix}\delta_0^2 = \max\left\{\delta_0^1, \delta_\pi^1, \delta_0^2, \delta_\pi^2\right\} \\ \text{or} \\ \delta_\pi^2 = \max\left\{\delta_0^1, \delta_\pi^1, \delta_0^2, \delta_\pi^2\right\}\end{matrix}\right\} \text{then} \left\{\begin{matrix} F_Q^1 = \left(\dfrac{\max\left\{\delta_0^1, \delta_\pi^1\right\}}{\max\left\{\delta_0^2, \delta_\pi^2\right\}}\right)^{\frac{3}{2}} \\ F_Q^2 = 1\end{matrix}\right\}\end{aligned} \tag{8}$$

3 Results and discussion

The maximum load is expressed as a function of the axial load-carrying capacity. This is done in order to represent graphically the values of F_A, F_R and M that cause permanent deformation in the most loaded ball, as detailed in [7]. We have:

$$Q_{MAX} = \frac{C_{0a}}{Z \sin\alpha} \tag{9}$$

When we substitute equation (9) into equation (6), we obtain:

$$\frac{F_A}{Q_{MAX} Z \sin\alpha} = \frac{F_A}{C_{0a}}$$
$$\frac{F_R}{Q_{MAX} Z \cos\alpha} = \frac{F_R}{C_{0a}} \tan\alpha \qquad (10)$$
$$\frac{M/d}{Q_{MAX} Z \sin\alpha} = \frac{M/d}{C_{0a}}$$

which can be seen as the coordinates of a point in a 3D diagram, with axes F_A/C_{0a}, $(F_R/C_{0a})\tan\alpha$ and $(M/d)/C_{0a}$. When we study different deformation fields defined by (A, R, M) according to equation (4) and solve equation (6) for each case, the final result is a cloud of points that define the acceptance surface.

It is difficult to represent the acceptance surface graphically, since it must be mapped with the parametric coordinates (A, R, M); each set of parametric coordinates results in another set in the coordinate system $(F_A/C_{0a}, F_R \tan\alpha/C_{0a}, (M/d)/C_{0a})$. A MAPLE application has been developed to assess the cloud of points with various levels of accuracy N in order to map the surface, as follows:

$$\left.\begin{matrix} A(i) = i \\ R(i) = i \\ M(i) = i \end{matrix}\right\} \quad i = -N \ldots N \qquad (11)$$

This is done in such a way that all possible combinations are chosen, resulting in $(2N + 1)^3$ points. Figure 3 shows a rendered triangulation of the cloud of points for $N = 20$ (68921 points).

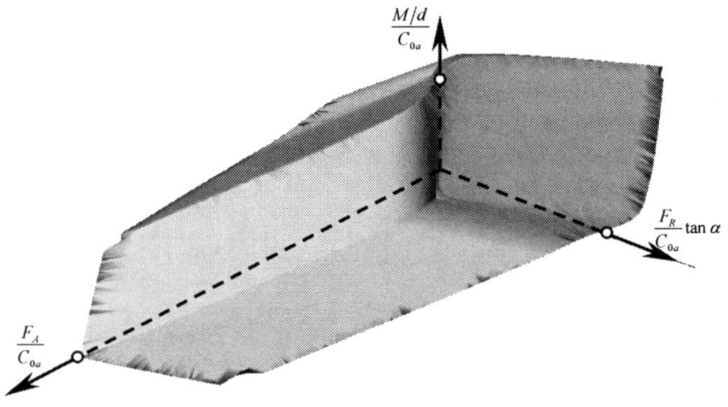

Figure 3: Moment–axial-force diagram for a four-contact-point.

The validity of the method has been verified by, for instance, checking the intersection of the surface with the coordinate axes. In fact, as can be found using the results in [10] and [11] applied to a four-contact-point slewing bearing, the limiting values of the axial and radial forces and tilting moment should be determined by

$$\begin{aligned} F_{A,\max} &= C_{0a} \\ F_{R,\max} &= \frac{2}{4.37} C_{0a} = 0.4577 C_{0a} \\ M_{\max} &= \frac{d}{4.37} C_{0a} = 0.2288 C_{0a} d \end{aligned} \quad (12)$$

These values match up with the ones given by the method developed in this paper. However, if the same comparison is done with intersection curves with the coordinate planes (as in [8] and [9]) it must be pointed out that there are some slight differences. In this sense, the method developed in this paper is more complete than those presented in [8] and [9], and it should not be seen as a superposition of the results of those methods. The reason is that in [8], only axial- and radial-force equilibrium is assumed, whereas the moment generated by axial non-uniform loading is not balanced, and in [9], only axial-force and tilting-moment equilibrium is assumed, whereas the axially generated radial force is not balanced.

4 Conclusions and future work

In this work a procedure for the assessment of the limiting values of the loads acting on a four-contact-point slewing bearing has been presented. The loads are an axial force, a tilting moment, and a radial force perpendicular to that moment. These limiting values are obtained by considering the equilibrium of the forces and moments in the inner ring, and then equating the maximum load to the value obtained from the axial load-carrying capacity. This procedure results in a surface expressed as a cloud of points in a three-dimensional coordinate system whose axes are the normalized loads. A designer can use this acceptance surface to provide a straightforward way to select a bearing.

Further work remains to be done, using for example multiparametric NURBS surfaces to assess equivalent loads in order to systematically obtain curves for the selection of bearings.

References

[1] Amasorrain, J.I., Sagartzazu, X., Damián, J., Load distribution in a four contact-point slewing bearing. Mechanism and Machine Theory, **38**, 479–496, 2003.

[2] Liao, N.T., Lin, J.F., A new method for the analysis of deformation and load in a ball bearing with variable contact angle. Journal of Mechanical Design, **123**, 304–312, 2001.

[3] Zupan, S., Prebil, I., Carrying angle and carrying capacity of a large single row ball bearing as a function of geometry parameters of the rolling contact

and the supporting structure stiffness. Mechanism and Machine Theory, **36**, 1087–1103, 2001.

[4] Antoine, J.F., Abba, G., Molinari, A., A new proposal for explicit angle calculation in angular contact ball bearing. Journal of Mechanical Design, **128**, 468–478, 2006.

[5] Hernot, X., Sartor, M., Guillot, J., Calculation of the stiffness matrix of angular contact ball bearings by using the analytical approach. Journal of Mechanical Design, **122**, 83–90, 2000.

[6] Jones, A., *Analysis of Stresses and Deflections*. New Departure Engineering Data, Bristol, CT, 1946.

[7] International Organization for Standardization, *Rolling Bearings – Static Load Ratings*, ISO 76:2006, 3rd edition, 2006.

[8] Sjoväll, H. The load distribution within ball and roller bearings under given external radial and axial load. Teknisk Tidskrift, Mek., h.9, 1933.

[9] Rumbarger, J., Thrust bearing with eccentric loads. Mach. Des., February, 1962.

[10] Harris, T.A., Kotzalas, M.N., *Rolling Bearing Analysis: Essential Concepts of Bearing Technology*. Taylor & Francis/CRC Press, 2007.

[11] Harris, T.A., Kotzalas, M.N., *Rolling Bearing Analysis: Advanced Concepts of Bearing Technology*. Taylor & Francis/CRC Press, 2007.

Gear micropitting: model and validation

J. A. Brandão[1], J. H. O. Seabra[2] & M. J. D. Castro[3]
[1]*INEGI—Instituto de Engenharia Mecânica e Gestão Industrial, Porto, Portugal*
[2]*FEUP—Faculdade de Engenharia da Universidade do Porto, Departamento de Engenharia Mecânica e Gestão Industrial, Porto, Portugal*
[3]*ISEP—Instituto Superior de Engenharia do Porto, Departamento de Engenharia Mecânica, Porto, Portugal*

Abstract

A numerical model for the prediction of surface initiated damage on gear tooth flanks (micropitting and mild wear) is presented. This model hinges on a model of the mixed film lubrication regime and on the application of the Dang Van high-cycle multi-axial fatigue criterion. The numerical model developed was applied to the prediction of micropitting and mass loss of gear teeth in an actual micropitting test.
Keywords: gears, surface fatigue, micropitting, numerical model, wear.

1 Introduction

Micropitting is best described as the appearance of minuscule cavities on the surface of a gear tooth, most often near and bellow the pitch line. They are thought to be caused by the propagation of micro-cracks originating from the tooth surface and swiftly reemerging so as to form a pit with a typical depth of 10 μm [1, 2]. Widespread micropitting severely increases the noise level of a gear box, and in some instances may sufficiently weaken the tooth subsurface for a whole layer to be removed at once, introducing a shape defect.

Because of the small size of the pits, the crack propagation time is correspondingly small and the fatigue life is dominated by the initiation time. Additionally, the fact that the phenomenon remains circumscribed within the first few tens of micrometres below the tooth surface renders it extremely sensitive to the contact

WIT Transactions on Engineering Sciences, Vol 66, © 2010 WIT Press
www.witpress.com, ISSN 1743-3533 (on-line)
doi:10.2495/TD100031

stress perturbations caused by the roughness of the surfaces and forces one to consider a lubrication model able to deal with such perturbations: a mixed film lubrication model.

For all these reasons, a numerical model was introduced that attempts to predict the onset of micropitting fatigue cracks on the surface of spur gear teeth by taking into account the geometry, roughness, residual stresses induced by manufacturing processes and the lubrication regime and by applying a fatigue initiation criterion, the Dang-van multi-axial fatigue initiation criterion.

This work is a revisitation of that previously presented in [3] and [4], where the model, with its underlying principles and methodology were put forward in great detail. The reader is invited to consult these papers if he finds the somewhat more sketchy presentation offered here insufficient. This revisitation is rendered opportune because in the time since the presentation of these works some new information came to light in [5] regarding the lubrication of the mineral oil used as an example in these works.

2 Numerical model

The overall workings of the models are shown schematically in figure 1. As can be observed, the model is comprised of two great blocks: one for the calculation of the elastic stresses at each instant of the meshing of the gears and another for the application of the Dang-Van high cycle fatigue initiation model to each point within the depth of 30 μm inside the tooth surface. It is not within the scope of this document to explain the model in detail, which is fully exposed in [4]. Instead, the readers attention is directed to several elements that must be provided if the model is to have adequate accuracy:

- the exact surface geometry, including its roughness;
- the manufacturing induced residual stresses;
- the boundary friction coefficient;
- a load sharing function that balances the load between the full film elasto-hydrodynamic (EHD) and the boundary lubrication cases;
- the rheological properties of the lubricant from which the full film EHD friction coefficient may be derived.

The first through third item are resolved through measurement. In particular, the third item is pointed to the reader's attention because it was the source of some uncertainty in [4] that [5] has dispelled. As for the remaining items, as well as the actual fatigue initiation criterion, they will be briefly discussed in what follows for the convenience of the reader.

2.1 Lubrication model

2.1.1 Contact pressure distribution

While in contact, a pair of meshing gear teeth is separated by a film of lubricant (most often, and the only case considered here, a lubricating oil). The film may separate completely the teeth surfaces, in which case the lubrication is full film

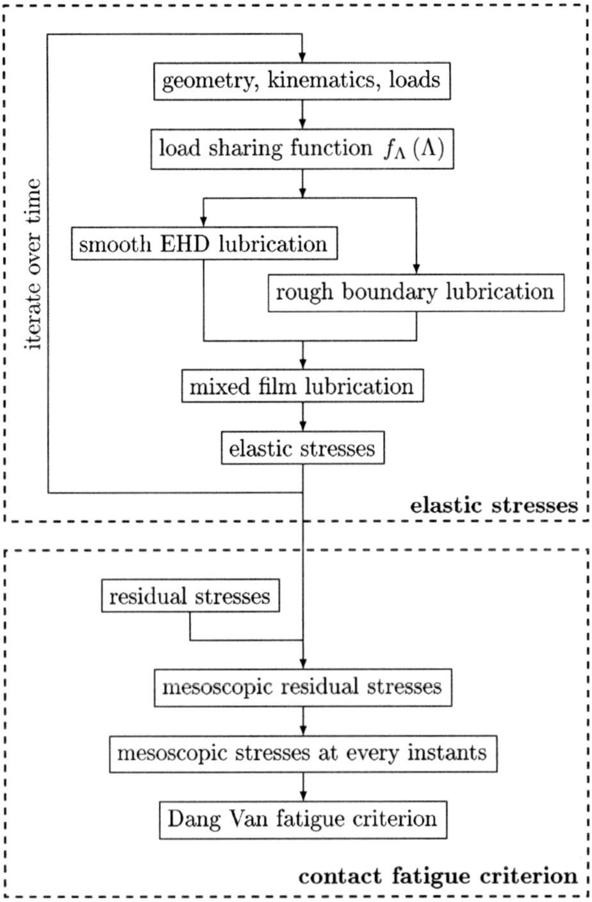

Figure 1: Diagram of the numerical model.

EHD (elasto-hydrodynamic); it may cease to bear load and only contribute to a decrease in the friction coefficient between the surfaces, in which case we have boundary lubrication; and it may be somewhere between those extremes, in which case we have mixed film lubrication, in which, while some separation of the surfaces is provided by the oil, direct contact still occurs between roughness features of the surfaces. It is generally acknowledged that the contact lubrication regime is controlled by the specific lubricant film thickness (Λ), calculated by:

$$\Lambda = \frac{\text{theoretical thickness of the film in EHD lubrication}}{\text{composite RMS roughness of the surfaces}} \tag{1}$$

When Λ is high ($\Lambda > 3$), the lubrication is full film elasto-hydrodynamic, when it is low ($\Lambda < 0.1$), we have boundary lubrication. Intermediate values define the range of mixed film lubrication.

The mixed film lubrication model proposed in [3] consists of a radical simplification whereby, instead of solving the full problem, a functions f_Λ of Λ is posited so that:

$$p = f_\Lambda(\Lambda)\, p^{\mathrm{EHD.T}} + [1 - f_\Lambda(\Lambda)]\, p^{\mathrm{BDR.T}} \tag{2}$$

in which, $p^{\mathrm{EHD.T}}$ is the theoretical contact pressure in full film EHD when ignoring the surface roughness, $p^{\mathrm{BDR.T}}$ the theoretical pressure in boundary lubrication when ignoring any bearing film and p the actual contact pressure.

2.1.2 Contact shear stress distribution

The shear stress within the contact is obtained as a sum of the contribution of the full film portion and the boundary portion.

It is generally accepted that in boundary lubrication the friction coefficient μ^{BDR} is essentially insensitive to variations of temperature, rolling velocity and slide-to-roll ratios and its value is to be found in a range approximately between 0.08 and 0.15. The actual values must be determined in each case of combination of lubricant and type of surface roughness with the aid of traction tests (see [5]).

All this means that the portion of the shear strain contributed by the boundary lubrication is:

$$\tau^{\mathrm{BDR}} = \mu^{\mathrm{BDR}}\,(1 - f_\Lambda)\, p^{\mathrm{BDR.T}} \tag{3}$$

In the case of the full film EHD contribution to the shear stress, things are more complex. Within a lubricant, its rheology establishes the relation of the shear stress with several other variables. In the present case the type of rheological law used is a $Bair - Winer$ visco-plastic law:

$$\dot{\gamma} = \frac{\tau}{\eta}\,\frac{-\ln(1 - |\tau/\tau_L|)}{|\tau/\tau_L|} \tag{4}$$

where τ is the shear stress, $\dot{\gamma}$ the shear strain rate, η is the low shear dynamic viscosity and τ_L, the limiting shear stress.

The low shear dynamic viscosity is here supposed to follow the Roelands viscosity law:

$$\ln\frac{\eta}{\eta_0} = (\ln\eta_0 + 9.67)\left\{\left(\frac{T - 138}{T_0 - 138}\right)^{-S_0}\left(1 + \frac{p}{196}\right)^{Z} - 1\right\} \tag{5}$$

where Z and S_0 are non-dimensional parameters of the oil, T_0 is a reference temperature in K, η_0 is the dynamic viscosity in Pa·s at the reference temperature an atmospheric pressure, T is the temperature of the lubricant in K, p is the pressure in GPa and η is the dynamic viscosity in Pa·s.

The limiting shear stress is supposed to follow a Barus type law:

$$\tau_L = \tau_{L0}\,\exp\left(\alpha_\tau p + \beta_\tau\left(T^{-1} - T_0^{-1}\right)\right) \tag{6}$$

where τ_{L0} is the limiting shear stress at atmospheric pressure and reference temperature T_0, α_τ and β_τ parameters to be determined empirically that account for the influence of pressure and temperature.

The pressure p is already known and equal to $(f_\Lambda \cdot p^{\text{EHD.T}})$. The strain rate is derived from the thickness of the theoretical full film EHD lubrication. The temperature is obtained by solving the thermal equation:

$$\rho C_p u \frac{\partial T}{\partial x} = k \frac{\partial^2 T}{\partial z^2} + \tau \cdot \dot{\gamma} \tag{7}$$

where ρ is the volumic mass, C_p the heat capacity, u the mean velocity and k the thermal conductivity.

The thermal equation and the rheological equation are coupled and, after having been simultaneously solved, yield the shear stress τ^{EHD} and the temperature T within the contact.

Finally, the actual shear stress is:

$$\tau = \tau^{\text{EHD}} + \tau^{\text{BDR}} \tag{8}$$

2.2 Dang-Van criterion

The Dang Van fatigue criterion [6] makes use of the *mesoscopic* stresses(stresses observed at the scale of the material grain) to predict the onset of high-cycle fatigue in a material point. These mesoscopic stresses are derived from the *macroscopic* ones (those obtained from ordinary continuum mechanics) as explained in [6, 3]. Knowing these mesoscopic stresses, the verification of initiation at any material point is expressed by the equation:

$$\tau_{\text{max}} + \alpha_{\text{DV}}\, p_H > \beta_{\text{DV}} \tag{9}$$

where τ_{max} is the maximum mesoscopic shear stress and p_H the hydrostatic stress (the negative of the hydrostatic pressure). The parameters α_{DV} and β_{DV} in equation (9) are fatigue properties of the material to be determined by performing a reversed torsion and alternating bending tests.

Whenever this expression is true, an initiation point is found. This is illustrated in figure 2, where the position of the mesoscopic stress state of a material point during a hypothetical load cycle on the p_H/τ_{max} plane is shown. It is seen that the path of the stress crosses the straight line delimiting the safety zone.

3 Gear micropitting: test and simulation

The numerical model's results are compared in this section to a gear micropitting test performed by Cardoso on an FZG test rig [7, 8]. The FZG type C gears used in the test were of DIN 20MnCr5 carburizing steel. The carburizing treatment left them with a 0.8 mm thick carburized layer and a surface hardness of 60 HRC. The gear oil is an ISO VG 150 paraffinic mineral oil that contains additives to increase its resistance to micropitting. Some of its properties are reproduced from [7, Table 1.1] in Table 1.

Table 2 lists the operating conditions for each load cycle. In addition the test procedure as well as the simulation procedure are described in Table 3.

Table 1: Gear oil properties.

Parameter	Method	Desig.	Value
Base oil	DIN 51451		Paraffinic mineral oil
Density at 15 °C	DIN 51757	ρ_{15}	0.894 g cm^{-3}
Kinematic Viscosity at 40 °C	DIN 51562	ν_{40}	153.6 cSt
Kinematic Viscosity at 100 °C	DIN 51562	ν_{100}	14.4 cSt
Viscosity Index	DIN ISO 2909	VI	96

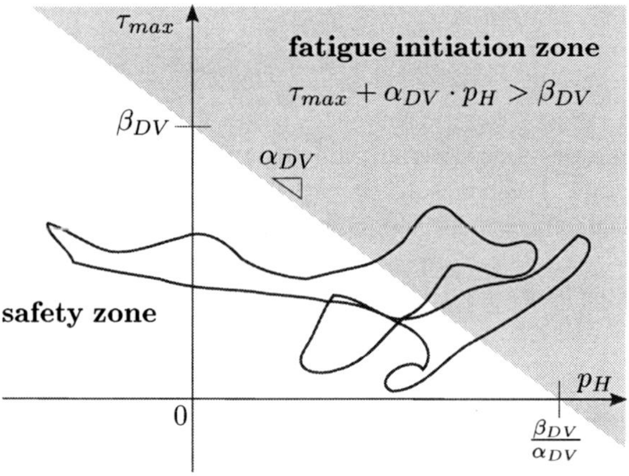

Figure 2: Position of the mesoscopic stress state of a material point during a load cycle on the $p_H/\tau_{\max}$ plane. The shaded half-plane represents the area where the Dang Van criterion is violated. Thus, the parts of the cycle placed in the shaded area violate the criterion.

Figure 3 shows the approximate positions where the tooth flank surface roughness measurements were made: on the left, those performed before load stage K3 and on the right those performed between successive load cycles (K6–K8 and K8–K9).

In addition to the parameters already presented, several others must be determined. In particular, the oil properties to be used in equations (5) and (6) must be known, as well as the boundary friction coefficient and the load sharing function.

The boundary friction coefficient of the oil was determined in [5] and was found to be approximately 0.09. The remaining parameters were obtained through an indirect method described in [4].

Table 2: Operating conditions of the micropitting test.

load stage	K3	K6	K8	K9
normal load F_N	851 N	2924 N	5073 N	6373 N
oil temperature	80 °C	90 °C		
duration	1 h	16 h		
number of cycles undergone by the driving gear	$135 \cdot 10^3$	$2.16 \cdot 10^6$		
angular velocity of the driving gear n_1	2250 RPM			
angular velocity of the driven gear n_2	1500 RPM			

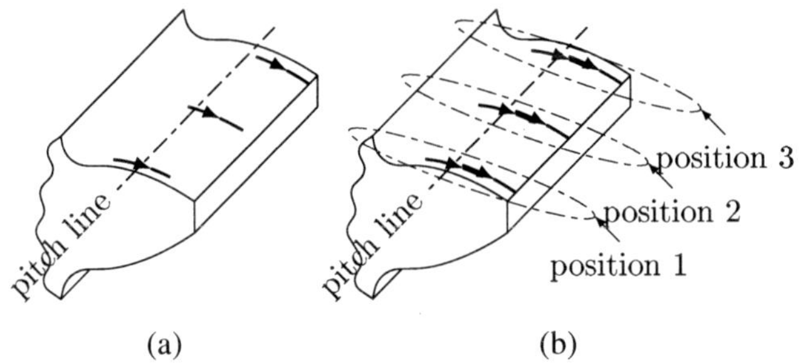

Figure 3: Positions of roughness measurement. a) the roughness measurements before stage K3. b) roughness measurements between load stages: each position includes one measurement below the pitch line and one above it.

For the application of the Dang Van criterion, its fatigue parameters α_{DV} and β_{DV} must be known as well. Of those, only the value of α_{DV} (0.987) was available in the open literature. Because of this each load stage was repeatedly simulated with different values of β_{DV} so as to scan the range 400 MPa $< \beta_{DV} <$ 900 MPa.

4 Comparison of the simulation with the gear test results

The test afforded two types of quantitative results: mass loss of the gears and roughness profiles at the end of each load stage. Neither is directly comparable to the types of results of the Dang-Van criterion which merely identifies *loci* of fatigue crack initiation.

A tentative solution was adopted here and will now be described. Consider figure 4, which shows the mechanism of crack propagation of a micropit proposed

Table 3: Gear micropitting test procedure and corresponding simulation procedure.

Test procedure	Step	Simulation procedure
Clean the gears in an ultrasonic bath, dry and weigh them.	1	
Measure the gear flanks roughness.	2	Introduce the roughness of the gears into the model.
Mount the gears on the test rig and perform load stage K3 during 1 h.	3	Run the simulation of load stage K3.
Without removing the gears from the test rig, take a lubricant oil sample.	4	
Submit the gears to load stage K6 during 16 h.	5	
Dismount the gears, clean them in an ultrasonic bath and dry them. Photograph the gear flanks surfaces. Weigh the gears.	6	
Repeat 2.	7	Repeat 2
Mount the gears on the test rig and perform load stage K8 during 16 h.	8	Run the simulation of load stage K8.
Repeat 6.	9	
Repeat 2.	10	Repeat 2
Mount the gears on the test rig and perform load stage K9 during 16 h.	11	Run the simulation of load stage K9.
Repeat 2.	12	Repeat 2
Measure the gear flanks roughness.	13	

in [1] in juxtaposition to an area of a tooth subsurface where a continuous patch of fatigue criterion violation is predicted. Both the dimensions and shapes in the sub-figures are similar. From here to assuming that they are essentially the same is a short step that the authors gladly take. Thus, in all that follows, whenever an area of fatigue initiation is found, it is assumed that the corresponding material will have been removed by the end of the load stage.

Table 5 shows the comparison of the measured mass loss in the driving gear at the end of each load stage with that predicted applying this propagation model considering two possible values of β_{DV}.

Even though there are no actual measurements against which to compare it, the value from the simulation of load stage K3 by itself is plausible, since it falls

Table 4: Parameters of the simulation.

Mixed and boundary lubrication parameters			
$f_\Lambda = \tanh\left(1.484 \cdot \Lambda^{0.258}\right)$		$\mu^{\mathrm{BDR}} = 0.09$	
Roelands viscosity parameters		**Limiting shear stress parameters**	
$T_0 = 363\,\mathrm{K}$	$\eta_0 = 1.56 \times 10^{-2}\,\mathrm{Pa{\cdot}s}$	$T_0 = 363\,\mathrm{K}$	$\tau_{L0} = 25\,\mathrm{MPa}$
$S_0 = 1.28$	$Z = 0.61$	$\alpha_\tau = 1.7\,\mathrm{GPa}^{-1}$	$\beta_\tau = 0\,\mathrm{K}$

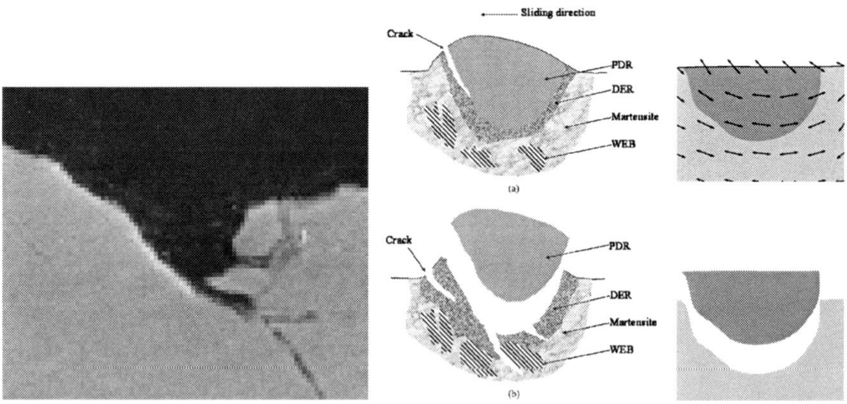

Figure 4: Approximate mechanism of crack propagation. (left) Photograph of the section of a micropit. (centre) Crack propagation mechanism according to Oila and Bull [1]. The crack propagates along the boundary between the plastic deformation region (PDR) and the dark etching region (DER). These, as well as the white etching band (WEB) are phases due to the change in steel micro-structure. (right) Approximate crack propagation mechanism adopted here: the crack propagates on the boundary of a Dang Van fatigue initiation patch until the patch is removed from the tooth.

well below the value measured at the end of the consecutive run of load stages K3 and K6.

The most dramatic fact to emerge from the data can be better observed with the help of figure 5, where the values of gear mass loss for the whole gear are also represented: the prediction of mass loss when $\beta_{\mathrm{DV}} = 400$ is nearly flawless. Not only that, but for other values of β_{DV}, the predicted mass loss curves stay resolutely parallel to the curve connecting the measured values. This advocates strongly for a very direct correlation of the Dang Van fatigue initiation area with the mass loss due to micropitting and, indeed, to a certain extent mild wear.

In figure 6 are shown as gray boxes the range of measured values of the roughness parameter R_{vk} below the pitch line of one tooth. Shown as well are

Table 5: Comparison of the measured mass loss at the end of each load stage with the predicted one.

	K3 [mg]	K3+K6 [mg]	K8 [mg]	K9 [mg]
model prediction with				
$\beta_{DV} = 500$ MPa	0.60	N/A	2.756	2.059
$\beta_{DV} = 400$ MPa	1.41	N/A	6.230	5.162
test measurements	N/A	3 ± 0.5	6 ± 0.5	5 ± 0.5

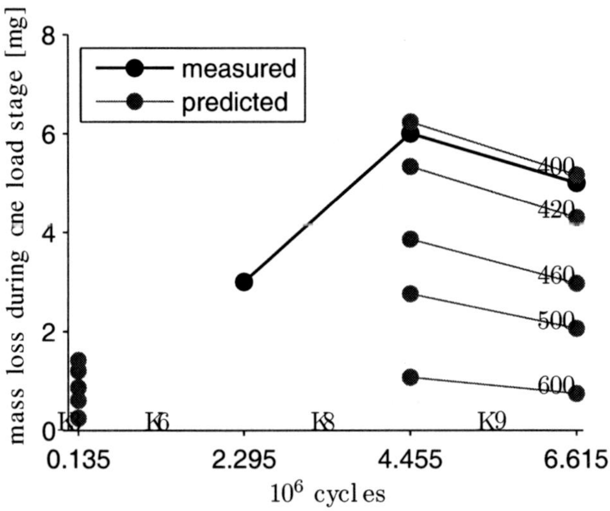

Figure 5: Mass loss of the driving gear in each stage: measured values and predicted ones for several simulation cases in which $\mu^{BDR} = 0.09$ and β_{DV} varies from 400 to 600 MPa.

the evolution of that same parameter as predicted by the proposed propagation model applied to several possible values of β_{DV}. Acceptable values of roughness parameters should fall within the range of measured roughness shown as shaded boxes. This is the case for the curves that match $\beta_{DV} > 500$. Unfortunately, those values also produce poor approximations to the mass loss. Once more "virtue is in the middle": there must be a value somewhere between 400 and 600 MPa that corresponds to the true β_{DV}.

However, some reflections with bearing on this quandary should be put forward. While the impression was given that the roughness measurement positions shown in figure 3 were precisely the same from load stage to load stage, this is hardly the

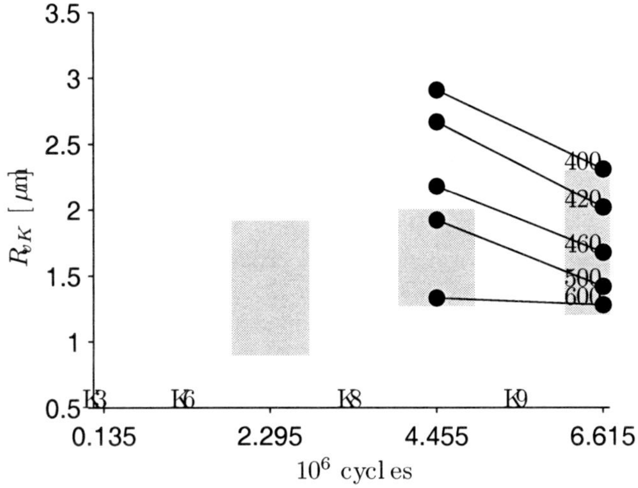

Figure 6: A tooth's evolution of roughness parameter R_{vk} below the pitch line. The shaded box show the range of the multiple measured values.

case: the operator performing the positioning of the gear at the beginning of each roughness measurement cannot be expected to accomplish this within a precision of a fraction of a millimetre. In practice, this means that an attempt is being made to uses one profile at a certain position as a starting point to predict another profile in a different position. On the other hand, this fact has much less bearing on the calculation of the mass loss, a global, integrated value much less sensitive to such slight variations.

5 Conclusion

A micropitting prediction model was applied in the simulation of an actual gear micropitting test. It has shown that an interval of values of β_{DV} can be identified within which the model is able to account for both the mass loss and, with some reservations, the roughness alterations on the surface of a gear. Thus, the carburized layer's fatigue property most needed for the application of the model is bracketed by values of approximately 400 MPa and 600 MPa. This goes a long way to confirm the validity of the model put forward here.

Acknowledgements

The authors wish to thank the "Fundação para a Ciência e Tecnologia" from the Portuguese Administration for its financial support to this work through the research contract "PTDC/EME–PME/66185/2006".

References

[1] Oila, A. & Bull, S., Phase transformations associated with micropitting in rolling/sliding contacts. *Journal of Materials Science*, **40(18)**, pp. 4767–4774(8), September 2005.

[2] Oila, A. & Bull, S.J., Assessment of the factors influencing micropitting in rolling/sliding contacts. *Wear*, **258(10)**, pp. 1510–1524, 2005.

[3] Brandão, J.A., Seabra, J.H.O. & Castro, J., Surface initiated tooth flank damage. part I: Numerical model. *Wear*, **268(1-2)**, pp. 1–12, 2010.

[4] Brandão, J.A., Seabra, J.H.O. & Castro, J., Surface initiated tooth flank damage. part II: Prediction of micropitting initiation and mass loss. *Wear*, **268(1-2)**, pp. 13–22, 2010.

[5] Brandão, J.A., Meheux, M., Ville, F., Seabra, J.H.O. & Castro, J., Experimental traction and stribeck curves of mineral, pao and ester based fully formulated gear oils. *IRF2009 – 3rd International Conference on Integrity, Reliability & Failure, Porto*, 2009.

[6] Dang Van, K., Griveau, B. & Message, O., On a new multiaxial fatigue limit criterion: theory and application. *Biaxial and Multiaxial Fatigue*, eds. M. Brown & K. Miller, EGF 3, Mechanical Engineering Publications, London, pp. 479–498, 1989.

[7] Cardoso, N., *Mechanical performance of biodegradable, low toxicity ester based industrial oils*. Master's thesis, Faculdade de Engenharia da Universidade do Porto, 2007.

[8] Martins, R. & Seabra, J., Gear micropitting performance of mineral and biodegradable ester gear oils. *Industrial Lubrication and Tribology*, **60(6)**, pp. 286–292, 2008.

A fast correction for traction-free surface of elastic quarter-space

R. Guilbault
Department of Mechanical Engineering,
École de technologie supérieure, Canada

Abstract

Applying the Hertz theory to some non-Hertzian contact problems can produce acceptable results. Nevertheless, including the influence of free surfaces requires numerical methods, many of which are based on the Boussinesq–Cerruti solution. This paper presents a new approach which is better capable of releasing quarter-space free surfaces from shear and normal internal stresses without any increase in calculation times. The mirrored pressure for shear correction is multiplied by a correction factor (ψ), which accounts for the normal load. The expression ψ is derived from the Hetényi correction process, and the resulting displacements show an enhanced correspondence with validation FEM models; with an imposed fluctuating pressure, the maximum edge displacement error was -21,90% for a shear load correction (Poisson coefficient $\nu = 0{,}3$), and introducing the ψ factor reduced the deviation to -9,55%, while for μ of 0,15, the maximum error was -11,30%, which was reduced to +0,60% with the ψ factor.

Keywords: *quarter-space, quarter-plane, free surface, contact model, Boussinesq–Cerruti solution, normal load.*

1 Introduction and literature survey

The Hertz theory can be applied to some non-Hertzian contact problems, and lead to acceptable results even when basic hypotheses are not fully respected. Such applications remain but approximated treatments, since the Hertz theory does not include the influence of free surfaces. In reality, including the influence of free surfaces requires numerical methods, the best-known being the finite element method (FEM). However, the computation time of FEM contact models can still have detrimental consequences. Fortunately, other accurate numerical

WIT Transactions on Engineering Sciences, Vol 66, © 2010 WIT Press
www.witpress.com, ISSN 1743-3533 (on-line)
doi:10.2495/TD100041

approaches exist, which are potentially more efficient in term of computation time.

Many 2D and 3D methods are based on the Boussinesq–Cerruti solution for point traction acting on elastic half-space (Hetényi [1, 2], Keer *et al* [3], Hartnett [4], Poon and Sayles [5], Li and Berger [6], and Chen and Wang [7]), where influence functions link surface tractions and displacements (Love [8], Li and Berger [6]). However, when used without correction in contact representations, such as in the Hartnett model (Hartnett [4]) these influence functions generate incorrect pressure increases in vicinity of free surfaces of real bodies. On the other hand, in contact problems with no free-surface effects, the Hartnett model produces very rapid and precise solutions. For this reason, many authors (Poon and Sayles [5], Chen and Wang [7]) have employed analogous modelling techniques for a variety of no-free-surface contact problems, and obtained accurate results.

The internal shear and normal stresses introduced by the Boussinesq–Cerruti-based methods on traction-free surfaces were already studied and solved by previous authors (Hetényi [1, 2], Keer *et al* [3]), and the free surface internal stresses are illustrated in fig. 1. The correction methods they proposed are exact and lead to a complete release of free surfaces. The technique requires repeated superpositions of the solution for the elastic half-space and then, significant computation efforts. Therefore, other authors (de Mul *et al* [9]) have observed that the shear load (τ) greatly influences the loaded surface displacement, while the normal load (σ) only plays a minor role. Consequently, as suggested by the Hetényi solution, a partial correction could be obtained when the pressure distribution is mirrored with respect to the free surfaces.

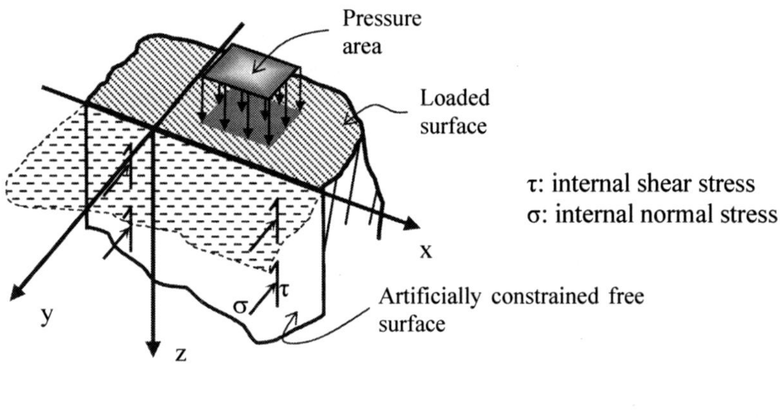

Figure 1: Half-space free surface internal stresses.

Since a partial correction does not noticeably increase the calculation time, yet offers visible gains in precision, this method has been successfully employed in machine element modelling (Guilbault *et al* [10]). However, in sensitive applications, the remaining σ could generate inadequate displacement calculation close to limiting edges of finite dimension bodies.

This paper presents a new approach which is better capable of releasing quarter-space free surfaces from τ and σ stresses without requiring increased calculation times as compared to the partial correction. The mirrored pressure for the τ correction is multiplied by a correction factor (ψ), which introduces an overcorrection to account for σ. The expression ψ is derived from the Hetényi correction process. Even though the proposed numerical approach is still an approximated correction, the resulting displacements show a good correspondence with FEM validation models.

2 Internal stresses

If a constant pressure (p_{ij}) is applied over a rectangular area (2a X 2b) of an elastic half-space, the normal displacement (δ_{kc}) of any surface point *kc* is calculated with eqns (1) and (2) (Love [8]). These equations are illustrated in fig. 2.

$$f_{kc,ij} = \left\{\frac{1-\nu^2}{\pi E}\right\}\left\{\begin{aligned}&(\xi_s+a)\ln\left[\frac{(\eta_s+b)+\left\{(\eta_s+b)^2+(\xi_s+a)^2\right\}^{1/2}}{(\eta_s-b)+\left\{(\eta_s-b)^2+(\xi_s+a)^2\right\}^{1/2}}\right]\\&(\eta_s+b)\ln\left[\frac{(\xi_s+a)+\left\{(\eta_s+b)^2+(\xi_s+a)^2\right\}^{1/2}}{(\xi_s-a)+\left\{(\eta_s+b)^2+(\xi_s-a)^2\right\}^{1/2}}\right]\\&(\xi_s-a)\ln\left[\frac{(\eta_s-b)+\left\{(\eta_s-b)^2+(\xi_s-a)^2\right\}^{1/2}}{(\eta_s+b)+\left\{(\eta_s+b)^2+(\xi_s-a)^2\right\}^{1/2}}\right]\\&(\eta_s-b)\ln\left[\frac{(\xi_s-a)+\left\{(\eta_s-b)^2+(\xi_s-a)^2\right\}^{1/2}}{(\xi_s+a)+\left\{(\eta_s-b)^2+(\xi_s+a)^2\right\}^{1/2}}\right]\end{aligned}\right\} \tag{1}$$

$$\delta_{kc} = p_{ij} \times f_{kc,ij} \tag{2}$$

where ξ and η are local coordinates, $\xi_s = \xi_{kc} - \xi_{ij}$ and $\eta_s = \eta_{kc} - \eta_{ij}$, ν = Poisson coefficient and E = Young's modulus.

Figure 2 also illustrates the τ elimination by the mirroring operation. In this representation, four correction mirrors are required because of the presence of two traction-free surfaces.

Hetényi presented a correction method for both the quarter-plane (Hetényi [1]) and the quarter-space problems (Hetényi [2]). After the mirror τ correction process, the σ distribution, which is actually doubled by the mirror

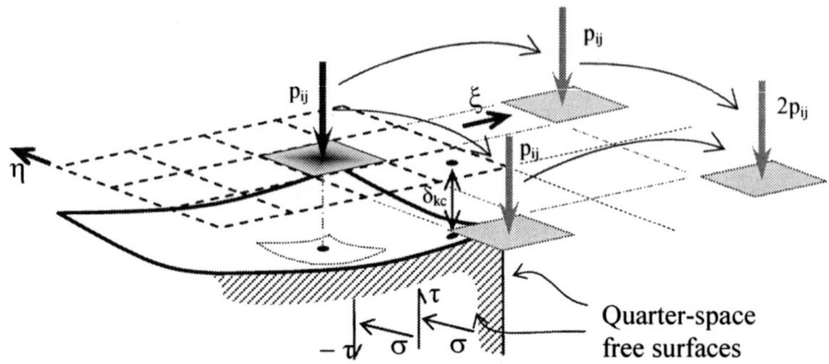

Figure 2: Pressure cells and mirror correction.

pressure application, is regarded as a distributed normal loading, and counterbalanced by a normal force distribution of the opposite sign.

The free surfaces must then be meshed, and the correction force distributions integrated into the problem by means of eqns (1) and (2). Since the correction forces generate new internal shear and normal stresses on the boundary, the mirroring process and normal forces elimination need to be repeated. The iterative procedure results in a complete correction of the free line or surface of the quarter-plane or space.

When considering the half-plane problem subjected to two concentrated forces P (fig. 3), the stress components at a given point (x,z) are given by eqns (3)

$$\sigma_x = -\frac{2P}{\pi}\left\{\frac{z(x+a)^2}{\left(z^2+(x+a)^2\right)^2}+\frac{z(x-a)^2}{\left(z^2+(x-a)^2\right)^2}\right\} \tag{3a}$$

$$\sigma_z = -\frac{2P}{\pi}\left\{\frac{z^3}{\left(z^2+(x+a)^2\right)^2}+\frac{z^3}{\left(z^2+(x-a)^2\right)^2}\right\} \tag{3b}$$

$$\tau_{xz} = -\frac{2P}{\pi}\left\{\frac{z^2(x+a)}{\left(z^2+(x+a)^2\right)^2}+\frac{z^2(x-a)}{\left(z^2+(x-a)^2\right)^2}\right\} \tag{3c}$$

At position $x = 0$ eqns (3) become

$$\sigma_x = -\frac{4P}{\pi}\left\{\frac{z(x+a)^2}{\left(z^2+(x+a)^2\right)^2}\right\} \tag{4a}$$

$$\sigma_z = -\frac{4P}{\pi}\left\{\frac{z^3}{\left(z^2+(x+a)^2\right)^2}\right\} \tag{4b}$$

$$\tau_{xz} = 0 \tag{4c}$$

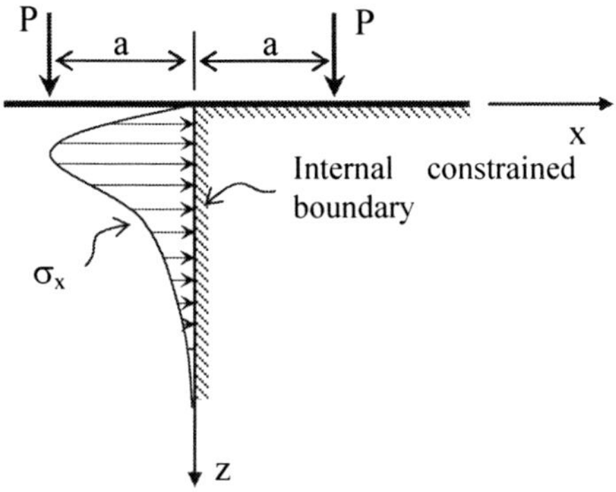

Figure 3: Half-plane with concentrated forces.

Integrating eqn (4a) with respect to *z* gives the total lateral force constraining the *z*-axis (eqn (5)).

$$\int_0^\infty \sigma_x dz = -2P\frac{1}{\pi} \tag{5}$$

Eqn (5) indicates that the resulting force amplitude is equal to the total applied load divided by π. Consequently, the elimination of this surface loading will, in turn, generate an additional normal load $(2/\pi)^2$P on the initial boundary, leading to geometric convergent series. Moreover, the line of action of the lateral force is determined as follows:

$$\frac{\int_0^\infty z\sigma_x dz}{\int_0^\infty \sigma_x dz} = \frac{\pi}{2}a \tag{6}$$

Eqn (6) shows that while the amplitude of the lateral force decreases, its position shifts away from the considered boundary. While the author did indeed establish the convergence of the process in a slightly different manner, this clearly illustrates the convergence of the Hetényi iterative process.

Hetényi ultimately formulated the total normal load acting on the initial half-plane boundary ($z = 0$) and on the boundary ($x = 0$) from eqns (7) and (8), respectively, where $[\sigma_z]_{z=0}$ and $[\sigma_x]_{x=0}$ are the cumulative stress distributions generated during the successive corrections.

$$C_z = \int_0^\infty [\sigma_z]_{z=0} dx = \frac{P}{a}\left(\frac{4}{\pi^2 - 4}\right) \tag{7}$$

$$C_x = \int_0^\infty [\sigma_x]_{x=0} dz = \frac{P}{a}\left(\frac{2\pi}{\pi^2 - 4}\right) \tag{8}$$

From eqns (7) and (8), Hetényi clearly demonstrated the complete elimination of the initial lateral load established with eqn (5). The inclusion of the third dimension in eqn (5) would still give the same lateral load amplitude (2P/π). Figure 4 illustrates the quarter-plane free boundary correction operation.

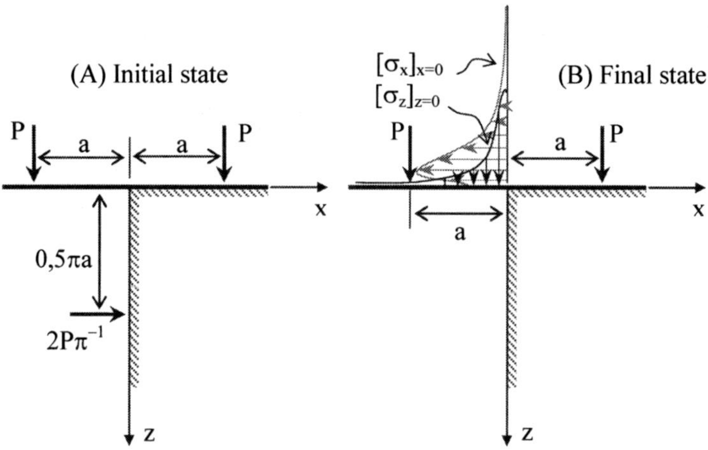

Figure 4: Quarter-plane states.

Since the objective of the present study is not to reproduce the Hetényi iterative process, which could be cumbersome, but rather, to compensate for the influence of the internal normal stress on the δ_{kc} amplitude, the following development of the factor ψ does not include the stress distributions, but rather, the corresponding concentrated forces.

3 Surface displacement

The displacements at any points in a solid subjected to a concentrated normal force *P* in the *z* direction are given by eqns (9).

$$u_x(P) = \frac{P(1+\nu)}{2\pi E}\left\{\frac{xz}{\rho^3} - (1-2\nu)\frac{x}{\rho(\rho+z)}\right\} \tag{9a}$$

$$u_y(P) = \frac{P(1+\nu)}{2\pi E}\left\{\frac{yz}{\rho^3} - (1-2\nu)\frac{y}{\rho(\rho+z)}\right\} \tag{9b}$$

$$u_z(P) = \frac{P(1+\nu)}{2\pi E}\left\{\frac{z^2}{\rho^3} + \frac{2(1-\nu)}{\rho}\right\} \tag{9c}$$

where $\rho = \sqrt{x^2 + y^2 + z^2}$.

If the quarter-space is reduced to a quarter-plane, fig. 5 illustrates the load conditions of the released half-plane (fig. (4(B))).

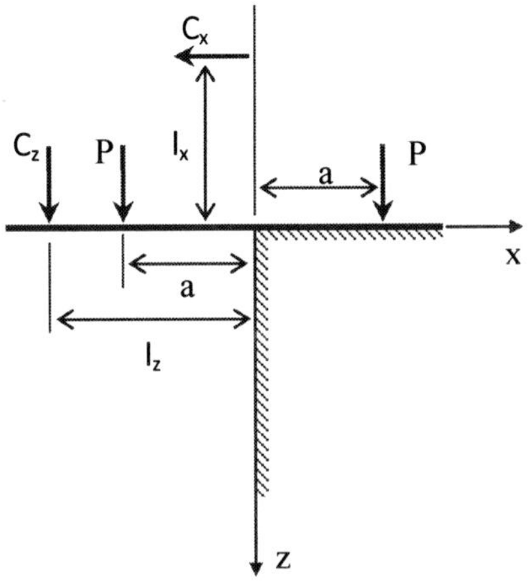

Figure 5: Quarter-plane conditions.

To quantify the influence of the correction forces (C_x and C_z) shown in fig. 5 on the z displacement at the point of application of P, eqns (9) can be rewritten as:

$$u_z(C_x) = \delta(C_x) = \frac{C_x(1+\nu)}{2\pi E}\left\{\frac{al_x}{\rho^3} - (1-2\nu)\frac{l_x}{\rho(\rho+a)}\right\} \tag{10a}$$

$$u_z(C_z) = \delta(C_z) = \frac{C_z(1+\nu)}{2\pi E}\left\{\frac{2(1-\nu)}{l_z + a}\right\} \tag{10b}$$

$$u_z(P) = \delta(P) = \frac{P(1+\nu)}{2\pi E}\left\{\frac{2(1-\nu)}{2a}\right\} \quad \text{(mirror force)} \tag{10c}$$

where $\rho = \sqrt{l_z^{\,2} + a^2}$, $l_x < 0$ and $C_x < 0$.

The cumulated influence of these corrective loads could be simulated by the mirroring process with an artificial overcorrection P* (eqn (11)).

$$\delta(P^*) = \delta(P) + \delta(C_x) + \delta(C_z) \tag{11}$$

This equation states that the elastic inherent rigidity of the half-space could be compensated through the addition of a displacement δ produced by a mirror corrective pressure P*. The P* amplitude must then be calculated with respect to the free surface in a given direction, allowing the direct transposition of the quarter-plane solution to the quarter-space problem. Replacing the real cumulative corrective load distributions with the equivalent forces (C_x and C_z) for speed gain will influence the final precision of the correction. That however, is an acceptable trade-off.

A single level ($n = 1$) of correction produces: $C_x = -(2/\pi)P$, $l_x = -(\pi/2)a$ and $C_z = (2/\pi)^2 P$, $l_z = (\pi/2)^2 a$. The corresponding correction factor ψ calculated form eqn (11) is then:

$$\psi = \frac{P^*}{P} = \left\{ 1 + \frac{2^5}{\pi^2(\pi^2+2^2)} + \frac{1}{(1-\nu)} \left[\frac{1}{\left(\left(\frac{\pi}{2}\right)^2+1\right)^{\frac{3}{2}}} - \frac{(1-2\nu)}{\left(\left(\frac{\pi}{2}\right)^2+1\right)^{\frac{1}{2}}\left(\left(\left(\frac{\pi}{2}\right)^2+1\right)^{\frac{1}{2}}+1\right)} \right] \right\} \tag{12}$$

This first stage correction reduces the initial lateral load $(2/\pi)P$ to $(2/\pi)^3P$ by the equivalent of 59,5%, whereas an $n \to \infty$ level of successive corrections theoretically lead to a complete release (eqn (13)). Eqn (13) is also a convergent series.

$$\psi = \frac{P^*}{P} = \left\{ \begin{array}{l} 1 + \displaystyle\sum_{n=1}^{\infty} \frac{2^{4n+1}}{\pi^{2n}\left(\pi^{2n}+2^{2n}\right)} + \\ \dfrac{1}{(1-\nu)} \left[\begin{array}{l} \displaystyle\sum_{n=1}^{\infty} \frac{1}{\left(\left(\frac{\pi}{2}\right)^{2(2n-1)}+1\right)^{\frac{3}{2}}} \\ -(1-2\nu)\displaystyle\sum_{n=1}^{\infty} \frac{1}{\left(\left(\frac{\pi}{2}\right)^{2(2n-1)}+1\right)^{\frac{1}{2}}\left(\left(\left(\frac{\pi}{2}\right)^{2(2n-1)}+1\right)^{\frac{1}{2}}+1\right)} \end{array} \right] \end{array} \right\} \tag{13}$$

Eqn (13) could be approximated by the following formulation.

$$\psi = \frac{P^*}{P} = 1{,}29 + \frac{1}{(1-\nu)}\left[0{,}17 - 0{,}25(1-2\nu)\right] \tag{14}$$

4 Validation

Eqn (14) is established to be integrated in a non-Hertzian contact model based on eqns (1) and (2). However, since the influence of the inherent normal stress can be minor in contact modeling, and consequently, veiled by the global precision. The validation models are built with predefined constant pressures applied over known areas, and three Poisson coefficient values ν are considered (0,15; 0,3 and 0,45). Figure 6 presents the tested form. The different cases are compared with FEM models, and to allow an isolation of the surface movement, the load is self-equilibrated.

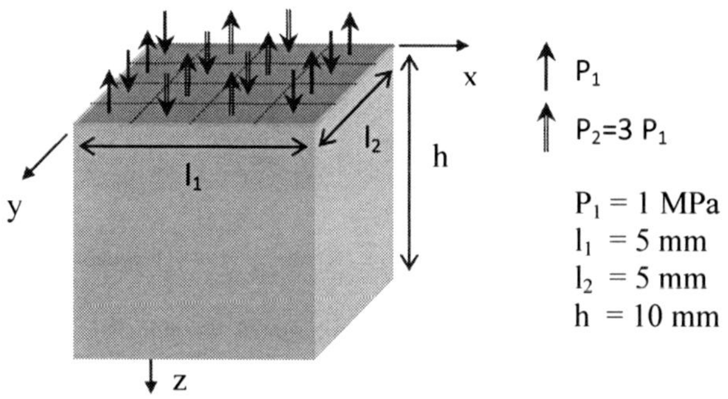

Figure 6: Test form.

The FEM models were meshed with parabolic tetrahedral elements of side length 0,125 mm through a 2,5 mm thickness (z direction). The remaining portion of the volume was meshed with gradual increasing element length (up to 1 mm).The form was also treated with constant pressure square areas (side length = 0,0625 mm), and the surface displacement was calculated with eqns (1) and (2), and corrected for internal stresses (τ and σ) through the mirroring process combined with the overcorrection factor ψ.

For the sake of clarity, 2D charts are preferred. Figures 7, 8 and 9 thus present the displacement values along the edges of the loaded surface (x and y =0). Displacement values for a shear correction alone (ψ = 1) are also added.

A visual inspection of the charts reveals a general enhanced correspondence between the overcorrected results and the FEM evaluations, as compared to the shear correction alone (ψ = 1). The figures also indicate that slight differences remain, and increase with ν. Figure 10 draws the maximum error curves for edges $y = 0$ and $x = 0$, respectively. These figures reveal that the edge displacement error (compared to FEM models) has an almost linear relation with the ν value.

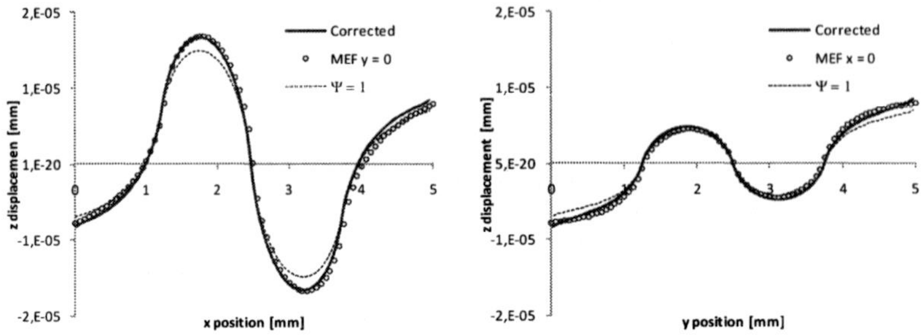

Figure 7: Edge displacement (ν = 0,15).

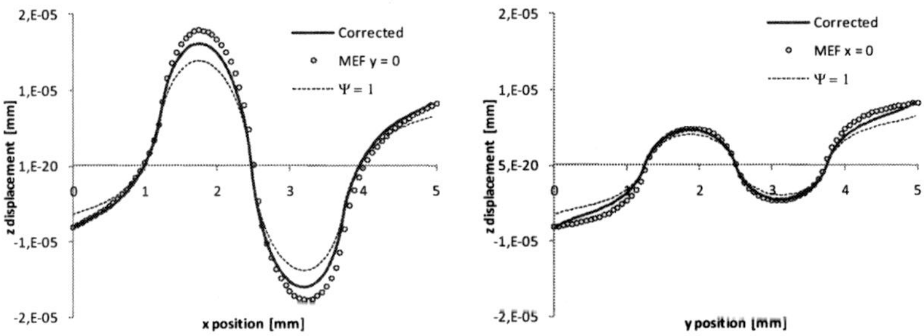

Figure 8: Edge displacement (ν = 0,30).

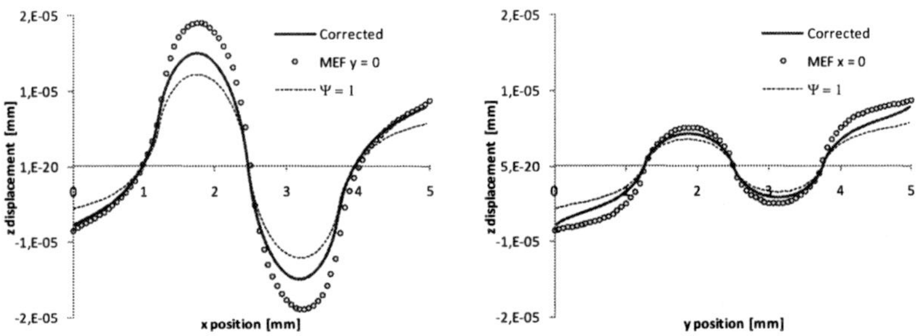

Figure 9: Edge displacement (ν = 0,45).

The maximum edge displacement error (along $y = 0$ edge) is -11,3% for a shear stress correction ($\psi = 1$) when $\nu = 0,15$, and introducing the ψ factor reduces the deviation to 0,60%. When $\nu = 0,30$ and 0,45, the maximum error amplitudes are -21,9% and -36,3% respectively, and are reduced to -9,55% and -21,1% with the ψ factor. This indicates that the ψ effectiveness varies with ν; for edge $y = 0$, the error reductions were 94,7% when $\nu = 0,15$, 56,4% when $\nu = 0,30$ and 42,0% when $\nu = 0,45$. This persistent difference is attributable to the concentrated equivalent forces C_x and C_z used in place of the real distributions.

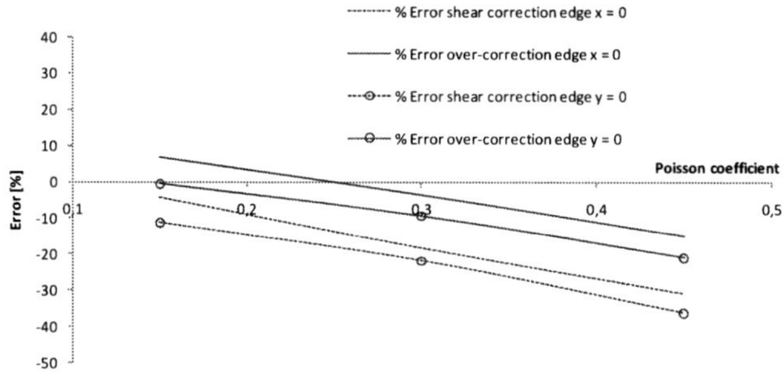

Figure 10: Maximum error curve along edge.

5 Conclusion

The Boussinesq and Cerruti non-Hertzian contact models offer precision and rapid solutions. However, the underlying half-space formulation introduces shear and normal stress distributions on traction-free surfaces. This paper presented the development of an overcorrection factor ψ accounting for the normal stress in the mirroring process. Compared to FEM models, it has been shown that the introduction of the ψ factor better simulates the elimination of the shear and normal stress influence on the displacement, and since no additional operations are necessary, the mirroring calculation times are not affected. The results reveal that the use of concentrated forces in place of real stress distributions causes deviations proportional to the Poisson coefficient of the material. When $\nu = 0{,}3$, the presence of ψ reduced the error on the edge from -21,9% for shear correction to only -9,55%, representing a 56,4% error reduction.

References

[1] Hetényi, M., A method of solution for the elastic quarter-plane. *J. of Appl. Mech.*, **82 (2)**, pp. 289-296, 1960.
[2] Hetényi, M., A general solution for the elastic quarter space. *J. of Appl. Mech.*, **37(1)**, pp. 70-76, 1970.
[3] Keer, L.M., Lee, J.C. & Mura, T., Hetényi's elastic Quarter space problem revisited. *Int. J. Solids Structures,* **19(6)**, pp. 497-508, 1983.
[4] Hartnett, M. J., A general numerical solution for elastic body contact problems. *Solid contact and lubrication,* eds H. S. Cheng & L. M. Keer AMD ASME, **39**, pp. 51-66, 1980.
[5] Poon, C. Y. & Sayles, R. S., Numerical Contact model of a smooth ball on an anisotropic rough surface. *ASME J. Tribol.,* **116(2)**, pp. 194-201, 1994.
[6] Li J. & Berger E.J., A Boussinesq-Cerruti solution set for constant and linear distribution of normal and tangential load over a triangular area. *Journal of elasticity*, **63(2)**, pp. 137-151, 2001.

[7] Chen W. W. & Wang, J. Q., A numerical static friction model for spherical contacts of rough surfaces, influence of load, material and roughness. *ASME J. Tribol.,* **131(2)**, 021402 8 pages, 2009.
[8] Love, A. E. H., The stress produced in a semi-infinite solid by pressure on part of the boundary. *Phil. Trans. Royal Society of London*, Series A, vol. 228, pp. 377-420, 1929.
[9] de Mul J. M., Kalker, J. J. & Fredriksson, B., The contact between arbitrarily curved bodies of finite dimensions. *ASME J. Tribol.,* 108, pp. 140-148, 1986.
[10] Guilbault, R., Gosselin, C., Cloutier, L., Express model for load sharing and stress analysis in helical gears. *ASME J. Mech. Des.,* **127(6)**, pp. 1161-1172, 2005.

Transient simulation of wear in a lobe pump using the wear processor

Y. J. Chen[1,2] & N. Huber[2,3]
[1]*Institut für Zuverlässigkeit von Bauteilen und Systemen, Universität Karlsruhe(TH), Germany*
[2]*Institut für Werkstoffphysik und Technologie, Technische Universität Hamburg-Harburg, Germany*
[3]*Institut für Werkstoffforschung, GKSS-Forschungszentrum Geesthacht GmbH, Germany*

Abstract

Lobe pumps represent an intelligent design principle for fluid technology. They combine competence in design and complex high-precision manufacturing processes and find their application in the fields of automotive and mobile hydraulics industry. The chambers for inlet, movement and outlet of the fluid are formed in a continuous process by rotating gear-shaped parts, which are sealed by several sliding contacts. The ability to predict wear in such complex systems is essential for the development and lifetime prediction. Using a dynamic 2D finite element simulation, the transient contact pressure and slip conditions are obtained on the surfaces of the three parts of which the pump consists: inner rotor, outer rotor and housing. For the wear simulation it is necessary to handle simultaneous wear in multiple contacts and contacting surfaces. It is shown that it is possible to predict wear in such a system over a large number of rotations with considerable effect on the geometry of the different components. The simulations allow one to predict the number of rotations for a given wear coefficient until critical gaps are formed, which can significantly reduce the performance of the lobe pump. Furthermore, the wear profiles show that the wear between the outer rotor surface and the housing reflects the pressure difference in the pump as well as the symmetry of the rotors. This work shows that it is possible to predict wear and the changing kinematics of systems, which are based on moving contacts. Problems of future interest are gears, camshaft follower systems, or injection systems.

Keywords: wear, wear prediction, transient contact, numerical simulation.

WIT Transactions on Engineering Sciences, Vol 66, © 2010 WIT Press
www.witpress.com, ISSN 1743-3533 (on-line)
doi:10.2495/TD100051

1 Introduction

Lobe pumps represent an intelligent design principle for fluid technology. They combine competences in design and complex high-precision manufacturing processes and find their application in the fields of automotive and mobile hydraulics industry. The pump design investigated in this work uses lobe rotors that turn continuously and has been fabricated in miniaturized dimensions [1]. An investigation with respect to short term fatigue and fracture of the lobe pumps has been carried out considering dynamic effects, friction, fabrication tolerances as well as braking and damping of fluid [2]. The time dependent forces and stress distributions in the components were found to be not critical with regard to failure due to short term fatigue or fracture. However, preliminary wear predictions under dry wear conditions indicated that the life time of the lobe pump might be short partially because severe wear occurs in the internal parts of the lobe pump [1, 3]. However, these results should be considered as a rough guess due to the complex transient behavior of multiple sliding contacts. Based on authors' best knowledge, there is still no literature available in dealing with the wear response of such systems and a realistic prediction of wear is not possible so far.

For wear simulation, Archard's wear model is the most popular model as discussed by many authors [4–14]. Archard's wear law represents a linear relationship between the incremental wear, local contact pressure and sliding distance. It was originally formulated on the global component scale in the form

$$\frac{V_W}{s} = k_D F_N , \tag{1}$$

where V_w is the wear volume, k_D is the dimensional wear coefficient, s is sliding distance and F_N is the applied normal force. To determine the local wear distribution in a tribosystem, commonly the local formulation is used,

$$\frac{h}{s} = k_D p , \tag{2}$$

where p is the contact pressure at a point on the surface. For a further generalization towards sliding distance dependent or time dependent wear prediction, eqn (2) needs to be formulated a differential equation [6]

$$\frac{dh}{ds} = k_D p , \tag{3}$$

with which an initial value problem can be solved through integration of the contact pressure over the slip distance. Therefore, the wear for each surface point results from the solution of the integral

$$h_w = k_D \int p ds = k_D \int p(t) \frac{ds(t)}{dt} dt . \tag{4}$$

In recent years, numerical simulation has achieved increasing attention as a powerful way for wear prediction and component design. As soon as multiple contacts and dynamic effects come into play, Finite Element simulations are so far the only way to achieve results, which account for the local contact conditions in the context of the system behavior. For well defined contact situations, for example, in gears, the global and local behavior has been studied with different multi-scale approaches [6, 11, 13].

In the case of the micro lobe pump, contacts show a wide variation in their extension from conformal to Hertzian type. Also the number of contacts in the system changes by opening and closing of contacts as a function of the rotation angle. These circumstances don't allow an application of efficient models, such as Winkler surfaces. Thus, the combination of a finite element simulation with wear prediction using the Wear-Processor are applied here and details are given in the next section. The wear predictions for a single wear step and for multiple wear steps as well as the gap and changing of the kinematics are presented in the third part, which is followed by the conclusions.

2 Finite element model

The finite element code ABAQUS [15] was used for the analysis of the micro lobe pump. The simulation with multiple sliding contacts was solved by the ABAQUS-Explicit, which has proven excellent in dealing with transient and nonlinear effects as well as finite rotations.

2.1 Geometry

The finite element models used in this paper are based on those used in [2]. For the completeness of this paper, the basic information of the finite element model is stated here.

The finite element model as shown in Figure 1 consists of four different parts: the housing, outer rotor, inner rotor and inner shaft. The outer diameter of the housing is 2.0 mm. These different parts are plugged together and, in this way, the different clearances can be taken into consideration. We assume zero initial clearance between the inner rotor and the inner shaft, while the clearances between the housing and the outer rotor and between the outer and inner rotor are non-zero, respectively (see Table 1). To study the influence of different clearances between the inner and outer rotor on the wear response, two different finite element meshes were adopted in this study.

Table 1: Contact definitions and variation of clearance.

Contact no.	Outerpart	Inner part	Clearance [μm]
1	Housing	Outer rotor	1
2	Outer rotor	Inner rotor	0, 1
3	Inner rotor	Inner shaft	0

The outer edge of the housing is fixed in all directions. The inner shaft is turned by a rigid surface, which is adjusted and tied to the inner circular edge of the inner shaft and which is connected to a reference node with only the rotational degree of freedom available. The simulation of this micro pump model

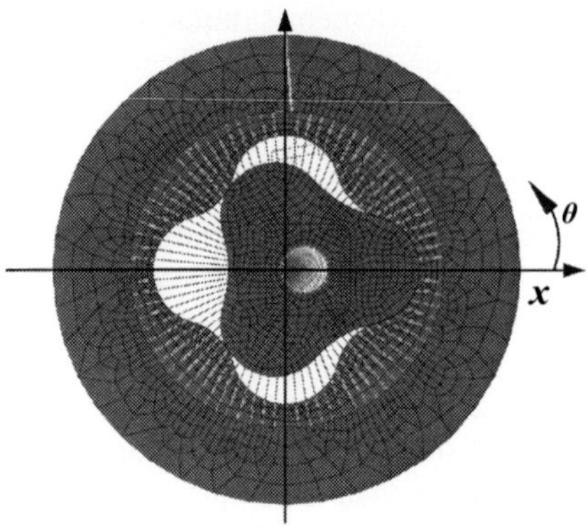

Figure 1: The finite element model of the micro pump.

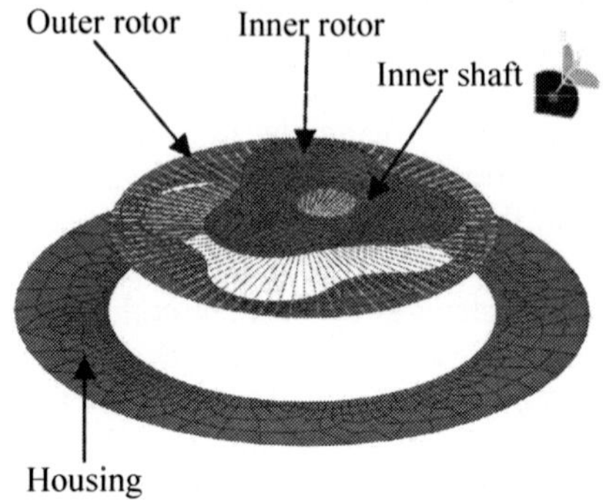

Figure 2: Detail configuration of the finite element model of the micro pump.

includes two steps. The first one is the acceleration step. The final angular velocity of 20,000 rpm at the inner shaft is applied with a constant acceleration when the inner shaft rotates to an angle of 120°. In the second step, the inner rotor rotates with the constant angular velocity of 20,000 rpm, and the simulation in this step is continued for 12 ms so that the inner rotor will rotate for 4 turns. The 4 turns ensure that the wear is the same with respect to the symmetry of the inner and outer rotors due to their 3 and 4 fold symmetry, respectively. The torque and force applied to this finite element model are the same as those in [2].

2.2 Material properties

In the finite element models, the housing and the rotors are made of ceramic while the inner shaft is made of steel. The behavior of the two materials is assumed to be elastic with the corresponding material constants given in **Error! Reference source not found.**. A Coulomb friction law with a friction coefficient μ=0.15 is applied to all contact surfaces. The wear coefficient has been chosen with k_D=13.5×10^{-8} mm^3/Nmm [13, 14].

Table 2: Material parameters.

Material	ρ [g/cm^3]	E [GPa]	ν [-]
Al_2O_3	3.99	400	0.23
Steel	7.80	210	0.3

2.3 Wear processor implementation

Based on the previous Wear-Processor 5.2 which has been described in detail in [8, 16], several changes have been implemented. Some are visible from the flow chart in Figure 3, the most important, relevant for this work, are explained below.

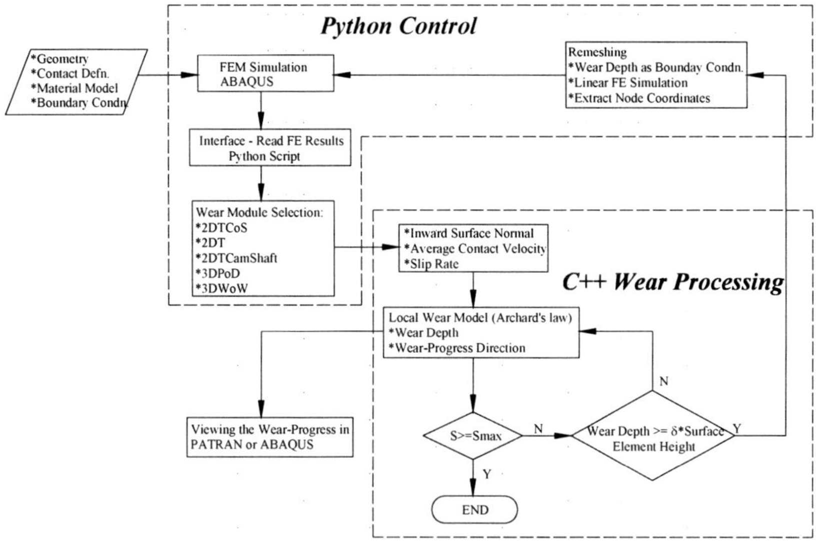

Figure 3: Flow chart for the working of Wear-Processor 6.0.

A) Contact variables

The latest version of ABAQUS 6.8.1 [15] provides the contact pressure p of the slave and master directly. Besides, in the ABAQUS-Explicit solver, the slip variables are provided for both the slave surface and master surface, thus, the mapping procedure of the slip rate from the slave surface to that of the master surface [8, 16] will not needed in the present case.

B) Python control

The flow chart can now be separated into two parts; the python control part and the C++ wear processing part. Preparation and starting of the contact simulation, reading FE results interface, wear module calling, and re-meshing process are integrated in the python control part. The C++ wear processing part of the code will carry out the wear calculation for each corresponding wear module. Applications with the respective modules have been published earlier for a 3D pin-on-disc [8, 16], a twin wheel tribometer [9] and a 2D gear tooth [13].

3 Results and discussions

3.1 Single wear step

The wear simulations are carried out for initial clearances between the outer and inner rotor of 0 and 1μm (see Table 1), respectively. In the contact simulation, the inner rotor is rotated for 4 turns, and thus the outer rotor will rotate for 3 turns because the transmission ratio between the inner rotor and the outer rotor is 3:4. Actually, the wear of the components of the micro pump is very small after one contact simulation, and it will not have a large effect for the following contact simulation since the geometry change is quite insignificant. Thus, in one wear step, we can calculate a certain amount of wear based on one contact simulation. In this work, the number of wear increments in one wear step is 50. Thus, one wear step assumes 200 rotations of the inner rotor.

The wear distribution of the inner rotor, outer rotor and the housing with and without initial clearance are shown in Figures 4–7.

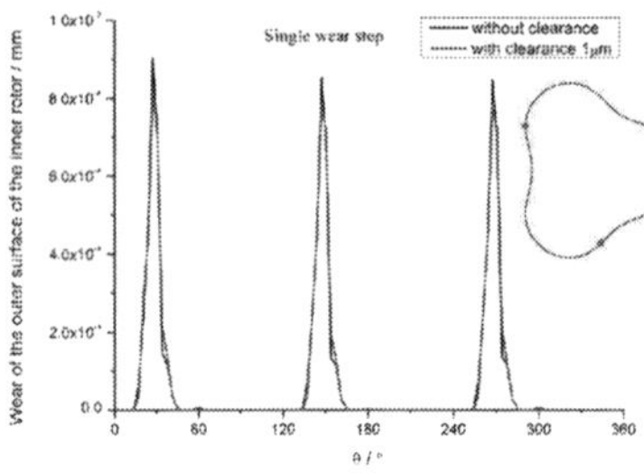

Figure 4: Wear of the outer surface of the inner rotor.

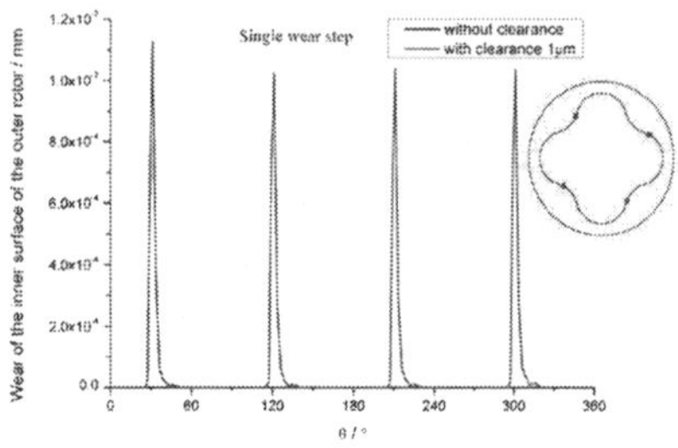

Figure 5: Wear of the inner surface of the outer rotor.

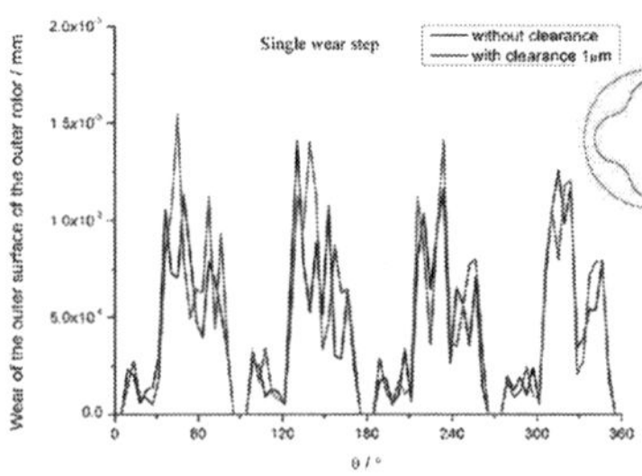

Figure 6: Wear of the outer surface of the outer rotor.

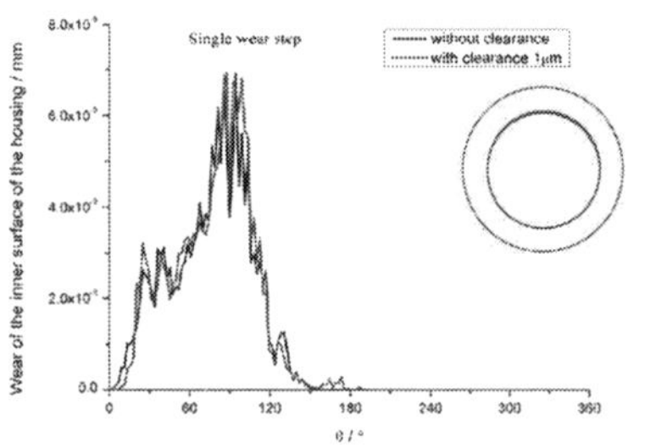

Figure 7: Wear of the inner surface of the housing.

Due to the geometric symmetry, a periodic behavior of the wear can be observed in the outer surface of the inner rotor as well as in the inner and outer surface of the outer rotor. The variable θ in the abscissa of these figures denotes the angle of each surface node with respect to the x axis in the counter-clockwise direction, as shown in Figure 1. The maximum wear of the outer surface of the inner rotor occurs at 27°, 147° and 267°, while the maximum wear of the inner surface of the outer rotor occurs at 30°, 120°, 210° and 300°.

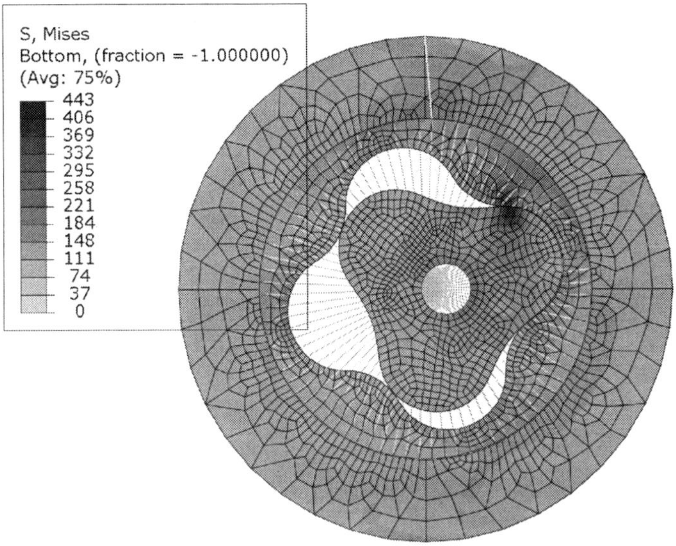

Figure 8: Von Mises stress distribution for turning angle θ=30.

These non-zero wear points are marked in the corresponding figure. Due to the upward applied force resulting from the pressure difference of the lobe pump, the maximum contact pressure always appears at the upper right contact region of the inner rotor and outer rotor contact pair, as shown in Figure 8. For the inner surface of the housing, only the top half of the surface experiences contact due to the upward force applied at the geometrical center of the outer rotor, and hence wear occurs only at this top region.

It can be seen from Figures 4–7 that the initial clearance between the inner rotor and outer rotor has little influence on the wear of the rotors and the housing. The wear in these figures is almost comparable in the respective whole surface.

3.2 Multiple wear steps

In this section, the simulation is carried out with the same parameters as those in section 3.1. The progress of wear of the inner rotor, outer rotor and the housing are shown in Figures 9–12.

As shown in Figures 9–11, the wear of the inner rotor and outer rotor approaches to a very stable periodic shape with the increasing of wear steps. The maximum amount of wear is exactly the same for the 3 teeth of the inner rotor, and also for the 4 teeth of the outer rotor. The worn areas of the outer surface of the inner rotor as well as the inner surface of the outer rotor increase with increasing number of wear steps. Furthermore, from these results it can be concluded that the initial clearance does have only little effect on the final wear distribution of the micro pump.

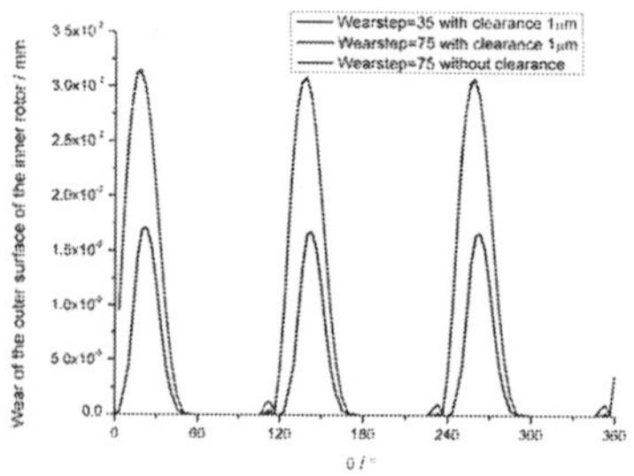

Figure 9: Wear of the outer surface of the inner rotor after multiple wear steps.

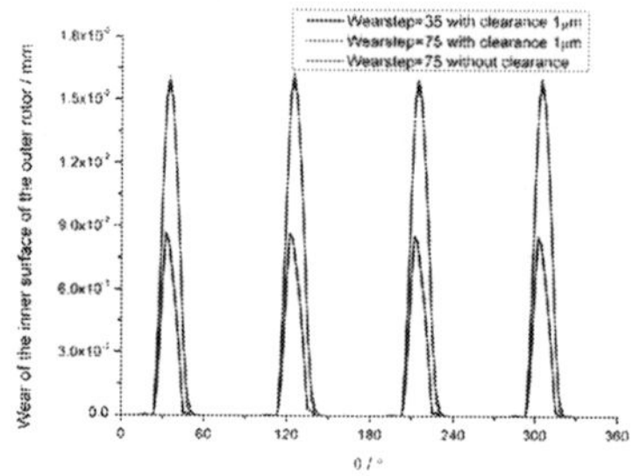

Figure 10: Wear of the inner surface of the outer rotor after multiple wear steps.

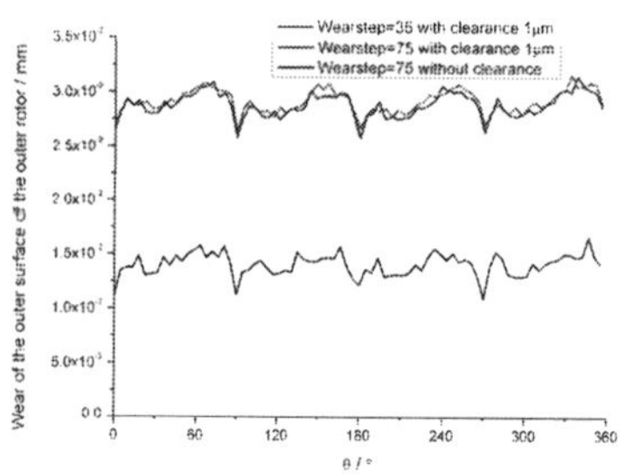

Figure 11: Wear of outer surface of the outer rotor after multiple wear steps.

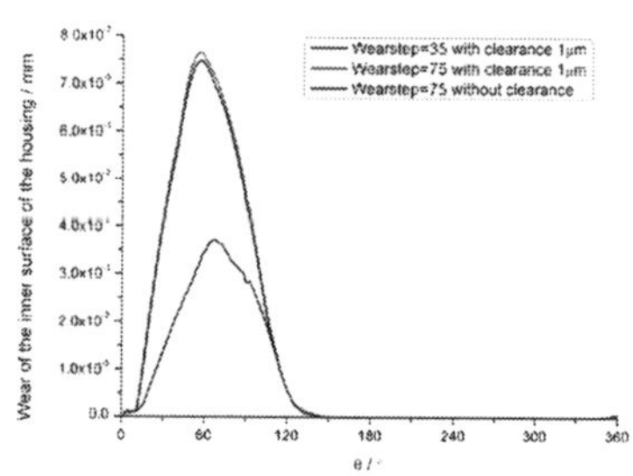

Figure 12: Wear of inner surface of the housing after multiple wear steps.

The wear distribution of the inner rotor, outer rotor and the housing show several interesting characteristics. After 75 wear steps, the maximum wear on the outer surface of the inner rotor is almost twice of that of the inner surface of the outer rotor. However, as we can see from Figures 4 and 5 for the first wear step, the maximum of wear of the outer surface of the inner rotor is smaller than that of the inner surface of the outer rotor. Thus, the rate of wear progress is not linear for the different contact surfaces. This indicates that a multiple wear step simulation is essential for the prediction of wear in such a complex dynamic system.

Furthermore, it can be seen from Figure 9 that the position of the maximum wear shifts left with increasing number of wear steps while the position of the maximum wear in shifts right. Also, the position of maximum wear of the inner surface of the housing in shifts left from θ=90° for the first wear step to 56° for the 75th wear step. These phenomena can be explained by the relative motions between the inner rotor, outer rotor and the housing. The outer surface of the housing is fixed in space, and the inner rotor which is controlled by a reference point can only rotate around with its geometric center. However, the outer rotor doesn't have a fixed rotation center, and therefore it can move dependent on the forming gaps due to wear. As a result, the center of the outer rotor gradually approaches to the upper right corner. As a consequence, the position of maximum contact pressure on the outer surface of the inner rotor moves to a smaller angle θ. For the position of maximum contact pressure on the inner surface of the housing we observe the same behavior. While for the outer rotor, it can be found that the position of the maximum contact pressure approaches a larger angle. Finally the change of the region with maximum contact pressure leads to the observed movement of the position of the maximum wear.

3.3 Gap formation and change of kinematics

The gaps between the inner rotor and the outer rotor (cf. Figures 8 and 16) as well as between the outer rotor and the housing increase with increasing number of wear steps, as shown in Figure 13. It can be seen that the gaps show a nearly linear increase. Using such a diagram, the maximum allowed number of rotations of the micro-pump can be predicted for a given maximum permitted gap.

Figure 14 shows the movement of the center of the outer rotor with increasing wear. Due to the upward force applied at the center of the outer rotor, the y-coordinate is always positive. In the very beginning, the outer rotor can rotate uniformly in the chamber and thus the x-movement in the horizontal direction is symmetric around the origin.

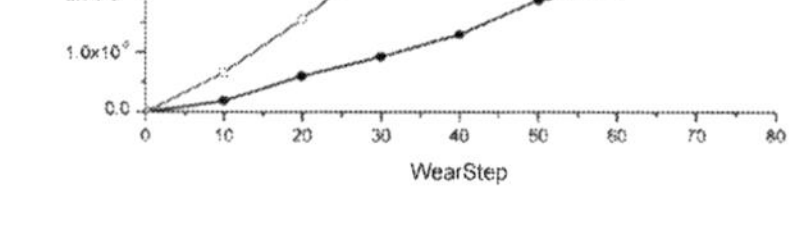

Figure 13: Gap formation with increasing number of wear steps.

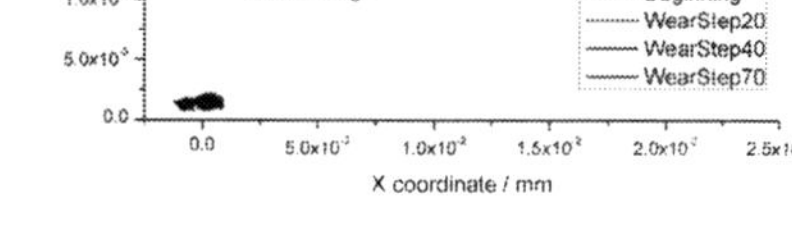

Figure 14: The movement of the center of the outer rotor with increasing wear step.

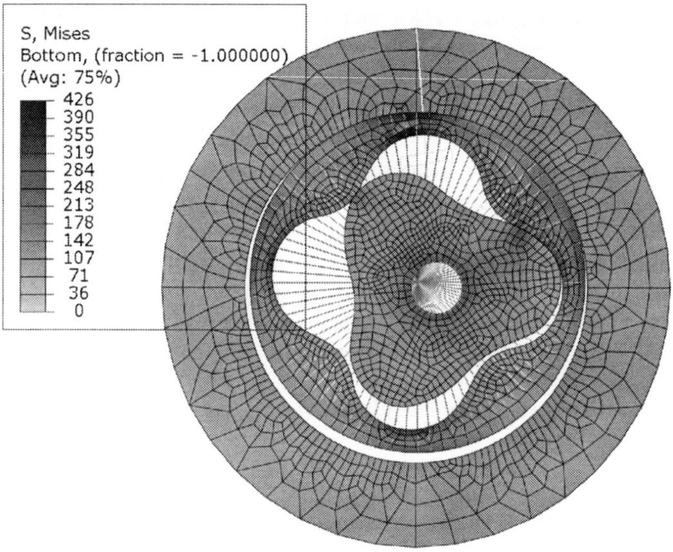

Figure 15: One main contact region after 40 wear steps.

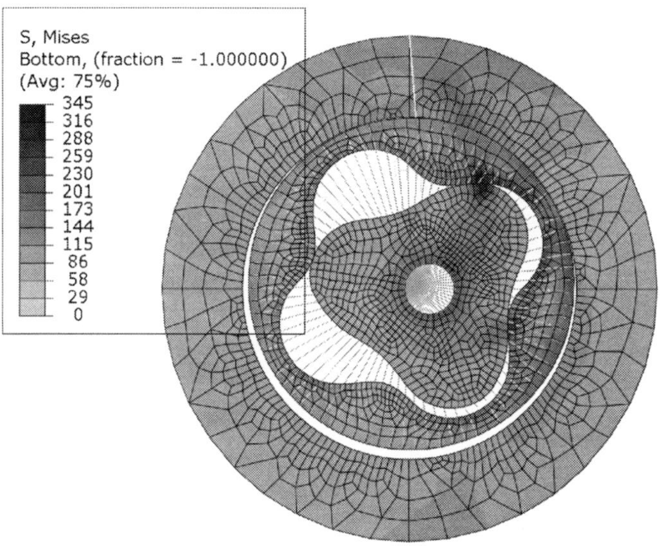

Figure 16: Two main contact regions after 40 wear steps.

With the occurrence of wear, the initial 4 point contact (see in [2]) is immediately reduced to a 2 to 3 point contact situation (see Figures 15 and 16 in which the stress distribution is given only at the contact surfaces). With the transition from a 2 point contact to a 3 point contact, the x-coordinate of the center of the outer rotor moves from the minimum to the maximum, as indicated in Figure 14.

With the increasing wear, it can be seen from Figure 14 that the center of the outer rotor gradually approaches to the upper right corner. This can be understood from the motion of the outer rotor by the action of inner rotor. Due to the special configuration of the inner rotor and outer rotor, the main contact regions between the two rotors always occurs at the right half side, which can be seen from with one main contact region and with two main contact regions at the right side. By combining the contact forces and the resulting pressure difference, the outer rotor approaches to the upper right position.

4 Conclusions

Based on the wear simulation tool Wear-Processor, the progress of wear in a micro lobe pump has been predicted. The simulation is based on the dynamic behavior of the pump. A single wear step wear simulation as well as a multiple wear step wear simulation was carried out. The results indicate that multiple wear step simulation is essential for correct prediction of wear in such a complex dynamic system. It has been shown that an initial clearance has little influence on the wear distribution on the various surfaces. The predicted gap formation depends nearly linear on the number of rotations.

With increasing wear, the center of the outer rotor moves to the upper right corner. The transition from a 2-point contact to a 3-point contact causes a shift of the x-coordinate of the center of the outer rotor from the minimum to the maximum.

The present work has successfully shown that wear can be predicted in complex dynamic systems. This opens up new possibilities for designers of engineering systems. In particular, it is possible to predict the effect of wear on the changing load, contact conditions, kinematics, and performance of the system under consideration. Furthermore, for given criteria, such as maximum gap size, amount of wear, maximum stress, etc, the simulation results provide a knowledge basis for the estimation of the life time in dependence of the component design and, in this way, can help to avoid expensive trial and error with prototype testing.

Acknowledgement

The authors would like to thank the German Research Foundation (DFG) for funding this work under transfer project T3 within the scope of the collaborative research center, SFB 499 – Design, production and quality assurance of molded micro-parts constructed from metals and ceramics.

References

[1] R. Ruprecht, T. Gietzelt, and J. Hausselt, Material and process development for wear loaded micro parts (in German). 2001, Forschungszentrum Karlsruhe: Karlsruhe, Germany. p. 47-54.

[2] N. Huber and J. Aktaa, Dynamic finite element analysis of a micro lobe pump. Microsystem Technologies, 2003, **9**(6-7): p. 465-469.

[3] M. Pfuff and W. Brocks, Simulation of contact and wear in ceramic microsystem parts (in German). 2001, Forschungszentrum Karlsruhe: Karlsruhe, Germany. p. 83-87.

[4] J.A. Williams, Friction and wear of rotating pivots in mems and other small scale devices. Wear, 2001, **251**: p. 965-972.

[5] W.G. Sawyer, Surface shape and contact pressure evolution in two component surfaces: Application to copper chemical mechanical polishing. Tribology letters, 2004, **17**: p. 139-145.

[6] J. Brauer and S. Andersson, Simulation of wear in gears with flank interference - a mixed FE and analytical approach. Wear, 2003, **254**: p. 1216-1232.

[7] K.E. Nürnberg, G. Nürnberg, M. Golle, et al., Simulation of wear on sheet metal forming tools - An energy approach. Wear, 2008, **265**(11-12): p. 1801-1807.

[8] V. Hegadekatte, N. Huber, and O. Kraft, Development of a simulation tool for wear in microsystems. Micro-engineering of Metals and Ceramics, Part II: Special Replication Techniques, Automation and Properties, ed. D. Löhe and J. Hausselt. Vol. 4. 2005: Wiley VCH, Weinheim.

[9] V. Hegadekatte, N. Huber, and O. Karft, Finite element based simulation of dry sliding wear. Modelling and Simulation in Materials Science and Engineering, 2005, **13**: p. 57-75.

[10] S. Dhanasekaran and R. Gnanamoorthy, Gear tooth wear in sintered spur gears under dry running conditions. Wear, 2008, **265**(1-2): p. 81-87.

[11] A. Fllodin and S. Andersson, Simulation of mild wear in helical gears. Wear, 2000, **241**(2): p. 123-128.

[12] H.L. Ding and A. Kahraman, Interactions between nonlinear spur gear dynamics and surface wear. Journal of Sound and Vibration, 2007, **307** (3-5): p. 662-679.

[13] V. Hegadekatte, J. Hilgert, O. Kraft, et al., Multi time scale simulation for wear prediction in micro-gears. Wear, 2010, **268**(1-2): p. 316-324.

[14] Y.J. Chen and N. Huber, Pressure-force transformation for transient wear simulation in two-dimensional sliding contacts. 2009, **submitted**.

[15] **ABAQUS**, Version 6.8.1. 2008, Dassault Systemes Simulia Corp.

[16] V. Hegadekatte, N. Huber, and O. Kraft, Modeling and simulation of wear in a pin on disc tribometer. Tribology letters, 2006, **24**(1): p. 51-60.

Section 2
Test methods

Fretting wear under magnetic field

Y. Nasu[1] & K. Sato[2]
[1]*Oyama National College of Technology, Japan*
[2]*Graduate School of Engineering, Chiba University, Japan*

Abstract

In our current society, based on science and technology, many kinds of electrical and mechanical systems are working under magnetic fields which become of a higher level year-over-year. The materials used in the above mentioned systems have suffered exposure to undefined risks due to magnetic effects. The authors have studied the complicated effect of magnetic fields on fretting wear, e.g. in some cases the effect is to accelerate wear volume and in other cases it does not. In the present paper the authors describe the outline of the effect of magnetic fields on fretting wear and the state of the art.
Keywords: magnetic field, fretting wear, friction and wear, material.

1 Introduction

Electrical and mechanical systems, such as electric generators, magnetic readers, linier motors, medical devices, and so on are working under the artificial magnetic field environment. According to recent progress in science and technology, devices and systems are increasingly exposed to and operate under higher magnetic fields year on year. The materials composed of the devices and systems, therefore, are exposed to an undefined risk situation, because the effects of magnetic fields have not been obviously examined. The problem is that the effects are different from conditions due to material, intensity of magnetic field, and working conditions of stress and vibration, and so on.

The wear of machine elements, or the contact surfaces, between them, is a reason for defining the lifetime of the electric apparatuses, such as conveyance equipment, rotating machinery, and precision electronics [1]. The wear of machine elements may be related to the intensity of the magnetic field and their lifetime is able to be controlled by changing the magnetic field. A number of practical examinations are, however, necessary for doing a good quality jobs in

WIT Transactions on Engineering Sciences, Vol 66, © 2010 WIT Press
www.witpress.com, ISSN 1743-3533 (on-line)
doi:10.2495/TD100061

design and maintenance. This is the motivation for our research through experiments to examine the effects of magnetic fields on fretting wear which is an especially important job in design and maintenance.

In the present paper, the effect of magnetic fields on wear and on friction aspects in the materials of the systems and devices has been reviewed. In addition, the result that the authors have studied of the effect of magnetic fields on fretting wear has been described. Furthermore, the relationship between the magnetic fields environment and advances based on science and technology has been surveyed.

2 Failure of urban systems under magnetic fields

Engineering structures and machines are frequently subjected to attacks due to corrosion, fatigue, and wear during working. As it is reported that the magnetic field affects these failures, deflection of the corrosion pattern [2] and improvement of the fatigue life [3], the authors summarize and focus on the effect of magnetic fields in wear as below.

2.1 Effects of magnetic field on wear process

Metal wear and wear process are affected by the kind and type of, direction of, and level of magnetic field, material, atmosphere, time, frequency, and so on.

2.1.1 Wear process under magnetic field

Muju and Ghosh [4], who firstly studied the effects of magnetic fields on wear, reported that magnetic fields affects wear volume in the state of adhesive wear. A standard wear test, with ferrous materials using the pin-on-disk method under vertical magnetic field applied by permanent magnets or electro-magnets, has been performed. The wear volume in ferromagnetic materials decreased under a magnetic field. This is expected to be related to that oxidation-promoting effect [5–7], hardness of frictional surface material [6], birth and death of wear particles at friction interface [7, 9], protection effect of oxidation film [7, 8] and acceleration of severe-mild wear transition [5]. Hiratsuka et al. [5], however, reported that wear volume increased under the horizontal magnetic field. Kumagai et al. [9] examined that wear volume decreases with decreasing magnetic field levels in both states, both sliding wear and reciprocating wear. The result shows that the magnetic effect in reciprocating wear is larger than that in sliding wear. The difference results from different behaviour in discharge of debris, or wear particles in wear process.

2.1.2 Combination of materials

In the magnetic field, material combination affects on wear volume that appears increase and decrease. For example, ferromagnetic materials become the mild-wear state and wear volume decreases. However, wear of one side may also increase a ferromagnetic material [10, 11]. This is associated to frictional surface, difference of the hardness, and embrittlement of the wear particles. On the one hand, the combination of paramagnetic and diamagnetic is not seen in the most of magnetic effects [12].

2.1.3 Effect of atmosphere

As wear volume increases under magnetic field which promotes oxidation, the wear behaviour in vacuum and in inert gases is interested. Kumagai et al. [13] performed experiments changing atmosphere pressure in a range of 10^5-10^0Pa. The magnetic field decreases wear volume in Ni/S45C, but an inflection point where wear changes from increase to decrease exists at a pressure of 10^3Pa in S45C/S45C: wear increased at pressure less than 10^3Pa and decreased at pressure greater than 10^3Pa. Zaidi et al. [6] also reports that wear of steel increases under magnetic field at low pressure of 5×10^{-5}Pa. It is reported for inert gas atmosphere that wear increased [14], and decreased [5] depending on material combination.

3 Effects of magnetic field on fretting wear

It has been reported by one of the authors that the wear volume of fretting wear is affected by the magnetic field [15, 16] and it increases under some cases of magnetic fields and decreases under other cases. As pointed out in the thesis of J.A. Collins, there are 55-influence factors in fretting, the fretting damage is very complicated. However, the influence of the magnetic field has not been clarified, and then the authors have worked to evaluate the effects by experiments.

3.1 Experimental details

In our research, the Herz contact between a ball and a plane has been applied. The detail of the experimental setup was introduced in our previous paper [17]. The fretting wear test device used is shown in Fig.1 (a). The small amplitude of repeated slip motion, at constant amplitude and frequency of sinusoidal wave, was applied using a piezo-electric actuator. The test specimens of SUJ2 steel ball (19.05mm dia.) and SPCC steel plate (W10×L10×T0.4 mm^3) were used. The surface roughness of steel ball and steel plate was R_{max}=0.25μm and 0.15μm, respectively. Vertical load was applied by dead weight. The vertical load in magnetic fields was applied by considering the value of magnetic attractive force between a ball and a plate. The tangential force was measured by a load cell using strain gage.

The magnetic field was applied to the vertical direction for a contact surface by electromagnet, as shown in Fig.1 (b). Magnetic field experiments were performed using an electromagnet at D.C. current of 1A or 3A. The magnetic density of near the contact surface was calculated using FEM analysis. The results of the analysis show the magnetic density of B=0.08T and 0.22T, respectively. The worn volume for fretting scar is modelled and evaluated geometrically. The diameter and depth of fretting annulus of the plate were measured with a laser microscope and calculated, as shown in Table 1.

The tests were carried out without lubricants, at a room temperature of 20-25°C, relative humidity of 70–80%, and in atmosphere for with and without magnetic field.

Table 1: Test conditions.

Magnetic density B[T]	Applied load P[N]	Fretting slip range D[μm]	Frequency f[Hz]	Number of cycle N
0.08 and 0.22	7, 20 and 46	6-38	10, 50, 100, 150	10^5-2×10^6

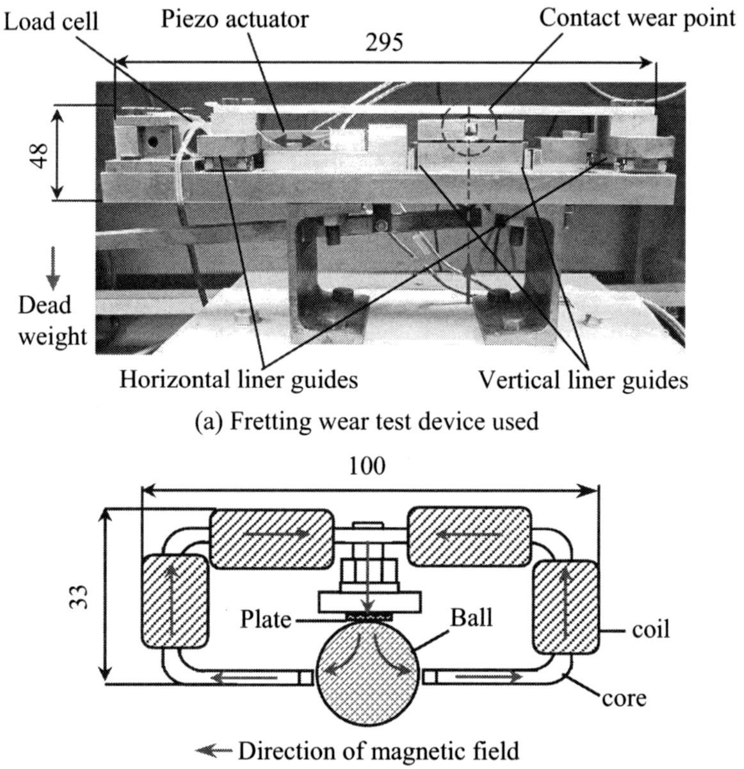

(a) Fretting wear test device used

(b) Closed up view of electro magnet used for tests under magnetic field

Figure 1: Fretting wear test rig used.

3.2 Experimental results

3.2.1 Effect of frequency

The appearance of worn surfaces fretted under different magnetic field densities is shown in Fig.2. The fretted annulus was spread in the case of (a) that is in lower frequency. The fretted annulus is broader at the low frequency of 10Hz,

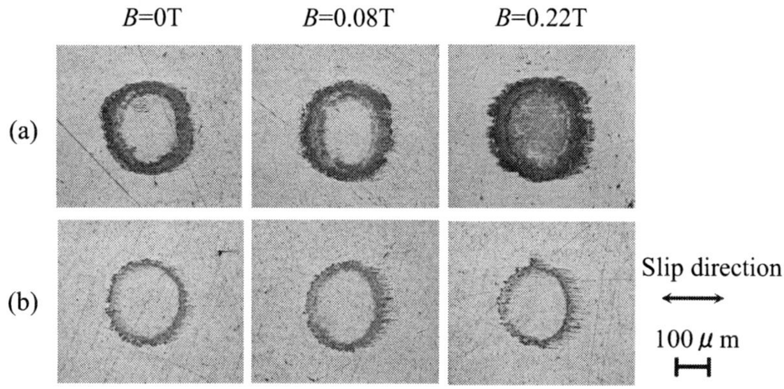

Figure 2: Effects of magnetic field and frequency of vibration on fretting wear annulus with and without magnetic field (P=20N, D=32μ m, $N=10^5$) (a) f=10Hz, (b) f=150Hz.

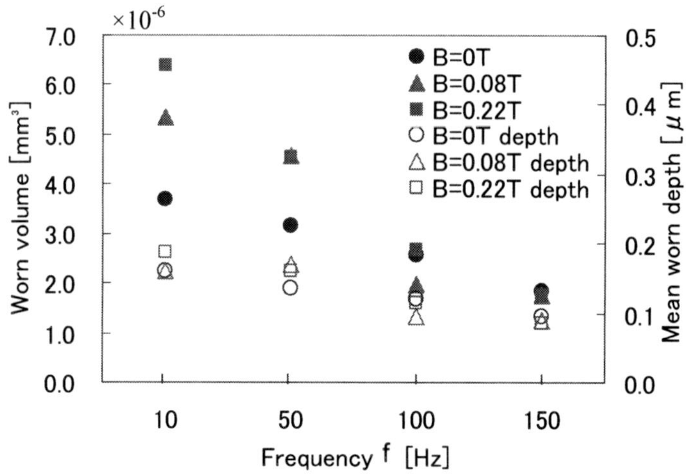

Figure 3: Effect of frequency on worn volume and depth (P=20N, D=32μ m, f=10Hz, $N=10^5$).

compared with at the high frequency of 150 Hz. The mean outer diameter of the annulus and worn volume are shown in Fig.3. The latter decreased with increasing frequency, and increased with increasing magnetic field density. That is due to mainly time dependency based on oxidation effects regarding to magnetic flux distribution in the contact area [16].

3.2.2 Effect of applied load

The slip motion is restrained by higher applied load, and then the state of contact becomes the Stick-Slip behavior. As a result, the influence of the magnetic field is decreased in the Stick-Slip region. At the high cycle region around 10^6 or

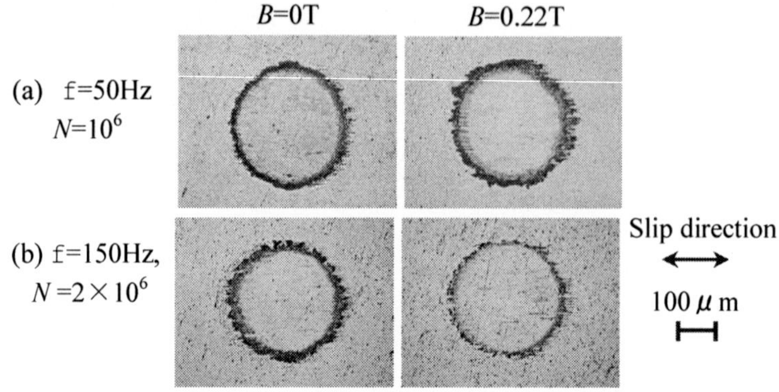

Figure 4: Effects of magnetic field and frequency of vibration on fretting wear annulus with and without magnetic field (P=46N, D=32μ m).

2×106 cycles, as shown in Fig.4. However, the influence of the magnetic field appeared similar to that at the lower applied load.

3.2.3 Effect of slip amplitude

The experimental result shows the effect of slip amplitude as follows: the wear morphology differs among Stick: Stick-Slip, and Slip regions, the increase in slip value is to clear form wear annulus and to be severe: the slip amplitude varied from the Stick-Slip region to the Slip region in the range of 32-38μm, and increases the discharge of wear particles.

Under the magnetic field, fretting scar changed when slip amplitude is large. The magnitude of slip amplitude appears in the change in the fretting scar. Fig.5 shows the relationship between two factors of b/a and Q/P, where b/a is the ratio of the fretting annulus diameter: b/a=1 is the Stick state, b/a=0 is the Slip state, and 0< b/a <1 is the Stick-Slip state. Experiment results are also plotted in Fig.5. Under the magnetic field, the values of b/a decrease with increase in Q/P. The decrease in the value of b/a in the magnetic field is greater in the range of Q/P=0.5-0.6, that is, the Stick-Slip region. The Stick region accelerates wear damage under the magnetic field (Fig.6). By the way, wear particles generated at the Slip region in the magnetic field are trapped, and therefore the worn volume is increased by the acts of abrasive particles which piled-up at the contact surface neighbourhood.

3.3 Accelerated oxidation under magnetic field

The worn volume is decreased by decompression, as results show in Fig.7 and the worn volume decrease to about 1/4. In vacuum (24Pa), worn volume does not change with and without magnetic field. In air, in the magnetic field the oxidation of the contact surface and wear particles are accelerated. That is considered to be resulting from reddish brown particles formed under the magnetic field. The authors, then, tried to measure the oxygen ingredient concentration of fretting scar about a contact surface in the slip direction by

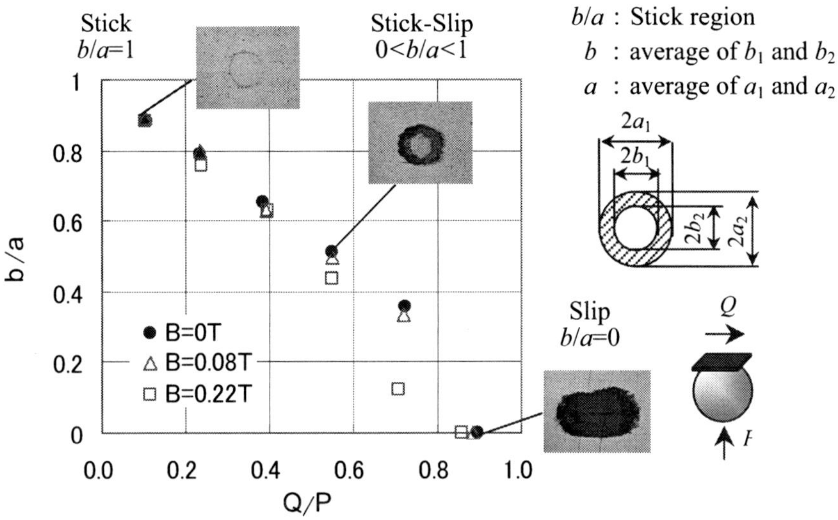

Figure 5: Relation between *Q*/*P* and *b*/*a*.

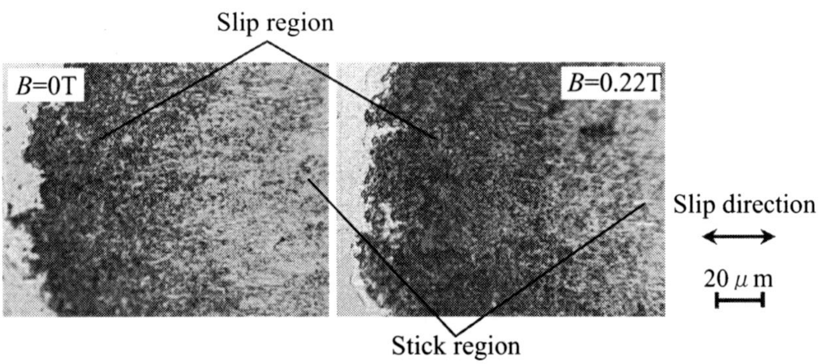

Figure 6: Laser micrographs of fretting wear scars on with and without magnetic field. (*D*=32μ m).

EPMA line analysis. The results are shown in Fig.8. Under the magnetic field, average oxygen concentration of worn scar was increased: In air is by 10% and in vacuum is 19.5%. In fretting wear process, the magnetic field promotes oxidation and proceeds to convert the reddish brown wear particles from black wear particles.

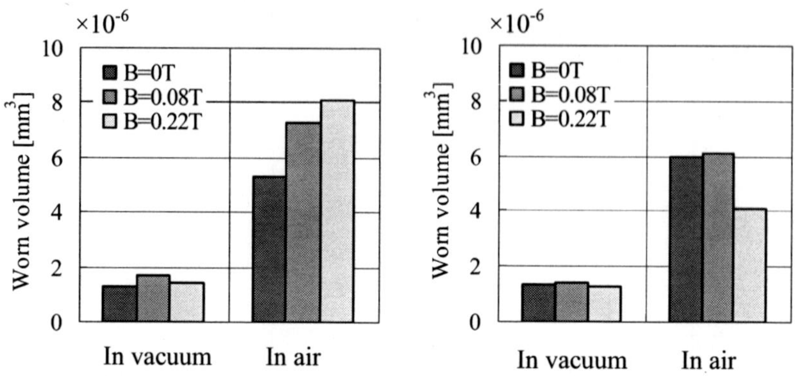

Figure 7: Relation between worn volume and air pressure. (a) f=10Hz, D=25μ m, N=105; (b) f=150Hz, D=32μ m, N=105.

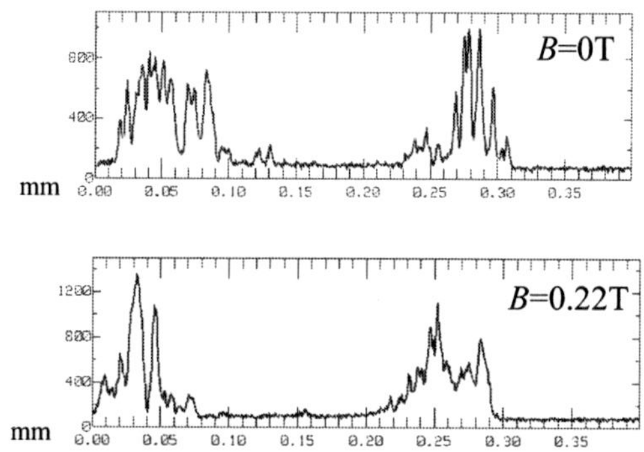

(a) In air.

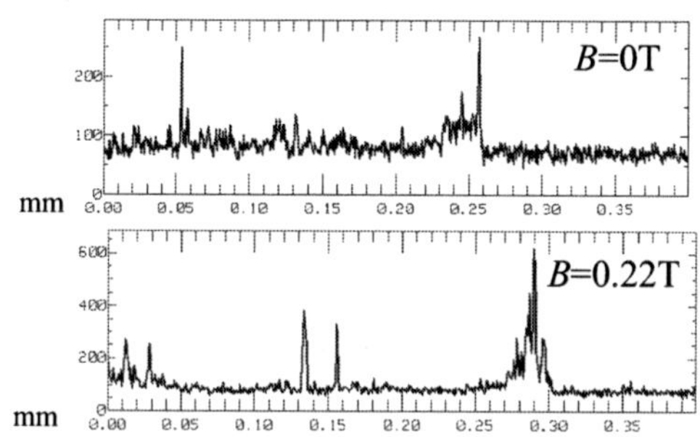

(b) In vacuum.

Figure 8: Oxygen analysis of worn surface by EPMA. (*P*=7N, *D*=25μm, *f*=10Hz).

3.4 Fretting map

The effect of magnetic field on wear morphology can be well understood through the fretting map showing the relationship between frequency and slip amplitude. That is shown in Fig.9. The effect of magnetic fields appeared only a boundary between the Stick-Slip and the Slip regions. The Slip region is reduced to slip amplitude at lower frequency by existence of a magnetic field, and the Stick-Slip region is shifted to high slip amplitude at higher frequency. In a long duration up to $N=10^5$ cycles, oxidization of the wear particles is accelerated in low frequency and they behave an abrasive material and act to decrease the Stick region.

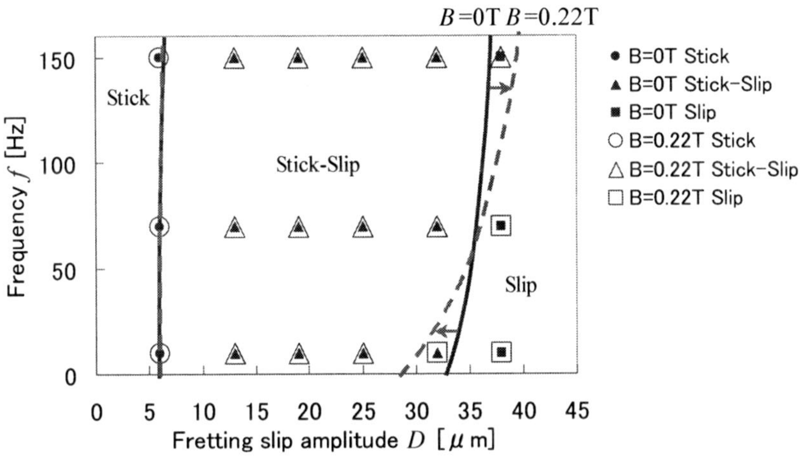

Figure 9: Fretting map with and without magnetic field (P=7N, $N=10^5$).

4 Conclusion

The effects of magnetic fields on machines have to be clarified for working under the magnetic fields environment. The effect of the magnetic field is very complicated; it is also known that there are many reports that the magnetic fields take place with opposite results. The experimental results showed that the wear region spread or narrowed under the magnetic field, as shown in the fretting map. This map is indicating the recommendation for the design of electric apparatus.

On the other hand, the applied technology of the magnetic field is reported. For example, in magnetic polishing in the processing [18], another, in the artificial joint in biotribology, problems such as the toxicity to the human body of the wear particles or the durability occur. Nakanishi et al. [19] applied a magnetic field to give an examination of the wear reduction, and therefore it is necessary to consider the influence on the human body and machinery under the magnetic field environment. In addition, with development of the machinery by superconductivity technology, mechanical behaviour of metallic materials under a magnetic field is examined [20]. However, in the superconductivity technology, the magnetic density is rising as shown in Fig.10 [21].

Therefore, we must do more research to clarify the influence of the magnetic field on machine and structures, because high technologies such as superconductivity technologies will be developed more rapidly.

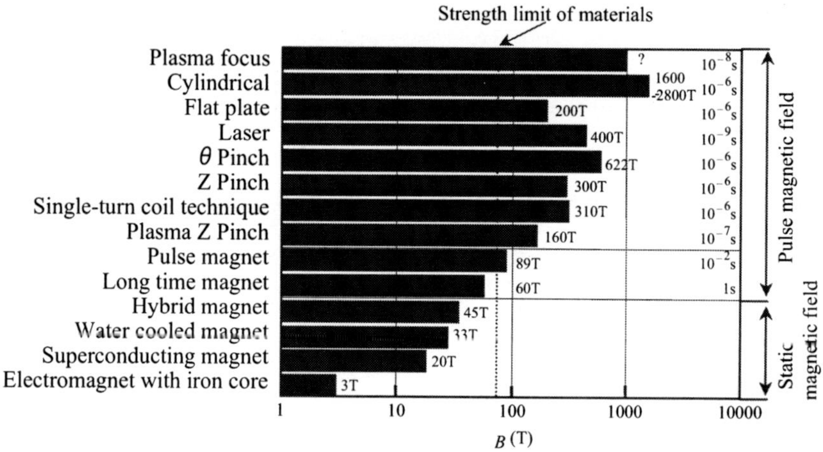

Figure 10: High magnetic field generation method and maximum magnetic field [21].

References

[1] K.Okada; Journal of Japanese Society of Tribologists, Vol.34, No.5 (1989) 379-380.

[2] A.Sato: Journal of Magnetics Society of Japan, Vol.21, No.4-2 (1997) 753-756.

[3] A.Celik, A. F.Yetim, A.Alsaran and M.Karakan, Materials and Design 26 (2005) 700-704.

[4] M.K.Muju and A.Ghosh, Wear, 41 (1977) 103-116.

[5] K.Hiratsuka, T.Sasada and S.Norose; Wear 110 (1986) 251-261.

[6] H.Zaidi, M.Amirat, J.Frene, T.Mathi and D.Paulmier; *Wear*, Vol.263, Issues 7-12, (2007) 1518-1526.

[7] K.Kumagai, O.Kamiya and S.Sato; Journal of Japanese Society of Tribologists; Vol.36, No.9 (1991) 727-734.

[8] H.Zaidi, K.J.Chin and J.Frene; Surface and Coatings Technology 148 (2001) 241-250.

[9] K.Kumagai, M.Mizui and O.Kamiya; Journal of Japanese Society of Tribologists, Vol.35, No.10 (1990) 741-748.

[10] M.El Mansori, H.Zaidi, E.K.Kadiri, and D.Paulmier; Surface and Coatings Technology 86-87 (1996) 511-515.
[11] H.zaidi, A.Senouci; Surface and Coatings Technology 120-121 (1999) 653-658.
[12] K.Hiratsuka, H.Saito and T.Sasada; Journal of Japanese Society of Tribologists, Vol.34, No.6 (1989) 430-436.
[13] K.Kumagai, H.Kawane and O.Kamiya; Scientific and Technical Reports of the Mining College, Akita University, 16, (1995) 51-58.
[14] M.El.Mansori, M.Schmitt and D.Paulmier; Surface and Coatings Technology 108-109 (1998) 479-483.
[15] K.Sato, T.A.Stolarski and Y.Iida; *Wear* Vol, 241 (2000) 99-108.
[16] Y.Iida, Y.Nasu and K.Sato; JSME Annual Meeting, Vol.2005, No4. 291-292.
[17] Y.Nasu and K.Sato; Transactions of the Japan Society of Mechanical Engineers, Vol.74, No.743, (2008) 1026-1030.
[18] O.Nakano; The Journal of the Surface Finishing Society of Japan, 57(11), (2006) 764-767
[19] Y. Nakanishi, T.Murakami and H. Higaki; Transaction of Japan Society of Mechanical Engineers. C, Vol. 164, No. 626 (1998) 3952-3958.
[20] E,Fukushima; Tetsu to Hagane, The Iron and Steel Institute of Japan, 75(6) (1989) 879-886
[21] N.Miura; Experimental Physical Chemistry series, Vol.5, Generation and Application of High Magnetic Fields (2008) KYORITSU SHUPPAN

Correlation between tribological measurements and surface characteristics of HVOF coated steel and PTFE materials in sliding contact

B. Fernandez-Diaz[1], M. Conte[1], A. Igartua[1], X. Fernandez[1], G. Mendoza[1] & A. Straub[2]
[1]*Manufacturing Processes Department, Tribology Unit, Tekniker, Spain*
[2]*Landing Gear Division, Liebherr, Aerospace Lindenberg Gmbh, Germany*

Abstract

In recent years, several improvements have been made concerning the wear of materials in sliding contact, thanks to the employment of Cr^{+6}-free hard coatings, their optimization, and the development of new deposition techniques and technologies. When solid lubricants, such as PTFE materials, are involved as sliding surfaces against steel material, roughness and hardness effects on friction and wear have to be taken into consideration, because they play an important role from the tribological stand point.

The aim of this work is correlation of tribological measurements with surface characteristics, in particular coating material, roughness and hardness of a landing gear rod sealing system. A cylinder-on-flat specimen configuration was chosen in order to faithfully reproduce working conditions, using a coated steel cylinder sliding against PTFE samples. The PTFE material was homogeneously worn independently of the countermaterial/coating, while when sliding against the softer material, the wear of the rod was observed. A comparison between results previously obtained on NBR is provided.
Keywords: friction, wear, PTFE, seals, Cr^{+6}-free, coatings, HVOF.

WIT Transactions on Engineering Sciences, Vol 66, © 2010 WIT Press
www.witpress.com, ISSN 1743-3533 (on-line)
doi:10.2495/TD100071

1 Introduction

Friction and wear reduction is one of the main objectives in sliding mechanical parts for minimizing loss of energy and improving systems performance. In recent years, attention has also grown in respect of maintenance cost savings, therefore a key question is how to achieve low levels of friction, as well as high wear resistance. In the field of elastomeric seals several approaches have been used, such as changing the system configuration, designing new cross section shapes [1–3], texturing the contact surface and applying coatings. Furthermore, in 1979 Gent et al. [5] studied the wear of metal by rubber attributing those phenomena at the direct attack upon metals of free radical species generated by mechanical rupture of elastomer molecules during abrasion. It is suggested that such studies might lead to new metal finishing processes and surface treatment that can have the double effect of improving the tribological performances and giving protection from external agents. Furthermore, coating technology is gaining ground thanks to new available technologies and focusing in particular on the need to use new alternative non toxic surface treatment with at least the equivalent functionality of Cr^{+6}. In a previous work the authors [4] tested the tribological behaviour of several Cr^{+6}-free hard coatings applied by HVOF (High Velocity Oxy-Fuel) sliding against polymeric samples, observing the strong influence of the cylindrical counterbody: the higher the cylinder hardness, the lower the wear resistance of the elastomeric samples. Evidences of abrasion, melting and adhesion were found with the hardest coatings. On the other hand, the rods with lower hardness were observed suffering the strongest wear when tested against the hardest rubbers. A deeper investigation is herein proposed, taking into account the previous researches carried out, considering the PTFE material sliding against the same materials and coating previously tested and the same working conditions in order to clarify the influence of the surface characteristics on the tribological measurements.

2 Experimental setup and specimens

In the test rig in Figure 1, a filled-PTFE sample obtained from a commercial seal used in actual application (1) is mounted in an AeroShell Fluid 41 oil bath (2) and loaded by an axially sliding rod (3), supported and fixed by two holders (4) on which a normal constant load is applied and transmitted to the axel of the rod. The holders and a couple of screws avoid rotation of the rod around its axel. The system is located in a climate chamber in order to set the temperature and humidity percentage. The normal load was set on 50 N for the firsts 30 s and then at 100 N, 25 Hz frequency and 2 mm stroke for each sample (see Figure 1). The coefficient of friction (CoF) was recorded at the regime condition for 30 minutes; each test was repeated 2-3 times.

The viscoelastic deformation and its effect on friction force were considered negligible in comparison to the other components due to the presence of fillers

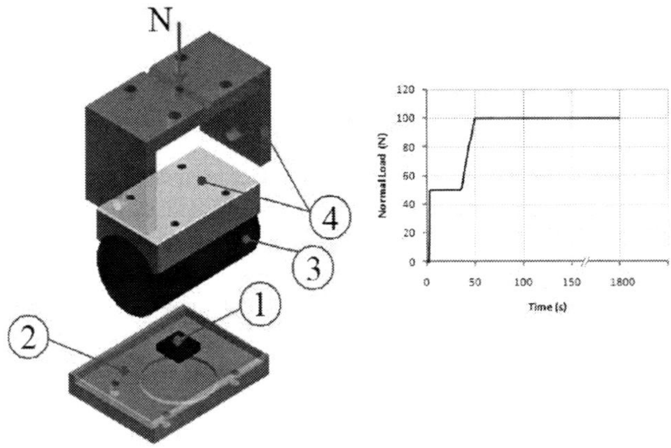

Figure 1: Test rig scheme and load history.

Table 1: Surface characteristics of coated rods.

Rod material and treatment	Coating method	Hardness (HV)	Ra (μm)
AlBr+G	HVOF	260	0,22
AlBr+G+F	HVOF	260	0,04
Uncoated+G	-	350	0,09
NiCrBSi+G	HVOF	745	0,16
NiCrBSi+G+F	HVOF	745	0,04
Cr+G	Electr. Bath	850	0,20
WCCoCr+G	HVOF	1115	0,23
WCCoCr+G+F	HVOF	1115	0,03
CrN+G	PVD	1974	0,19
ZrN/ZrCN+G	PVD	1975	0,16

(such as glass fibres, graphite and bronze), which are able to improve the mechanical characteristics of PTFE. Furthermore, no appreciable plastic deformation was observed.

Three different technologies were used for coating the steel rods: High Velocity Oxy-Fuel (HVOF), Electrolytic Bath and Physical Vapour Deposition (PVD). Subsequently the coated rods were exposed to different surface modification processes (G = Grinding, F = Superfinishing) to vary their texture. Vickers Hardness and roughness (Ra) have been measured and reported in Table 1.

The shot peening process was also applied to the rod with the aim of changing the texture of the surface and not to produce a compressive residual stress layer, modifying the mechanical characteristics of the metal. It was associated with grinding process to reduce the amount of induced-roughness peaks, but no relevant results were found. Therefore, they were not reported.

3 Results and discussion

Taking into consideration as reference the uncoated and Chromium coated by electrolytic bath rods CoF, applying grinding surface treatment to the rods coated by means of the HVOF process did not result in a clearly defined trend at any level of hardness (Figure 2). Uncoated material with grinding resulted from having better roughness (Ra) than the coated and grinded ones; therefore it was decided to also apply a finishing process to the coated cylinders, obtaining lower values of Ra and subsequently relevant improvement in terms of CoF, especially for the hardest coating. The shot peening process seemed to have no notable effects on the CoF. In general, the grinded+finished rods improve their CoF with hardness, while grinding alone had no relevant effect. PVD coatings of CrN and ZrN/ZrCN were also tested, also showing considerable improvements in comparison with the results obtained by the authors [4] on NBR specimens.

The effect of finishing treatment on the CoF is the opposite of that seen by the authors for NBR materials [3], independently of rod hardness, demonstrating that NBR has less ability to form a transfer film on the counterface, resulting in friction force increment (Figure 3). The CoF is more influenced then by roughness than by PTFE. A relevant difference between NBR rubber and PTFE can be observed in terms of friction at the initial stage of the test, when the high degree of lateral mobility of PTFE chains quickly induces molecular orientation in the sliding direction. This phenomenon is also aided by PTFE's smooth linear molecular profile and the ability of the electron clouds of the large fluorine ions to screen the bonding forces between neighbouring molecules.

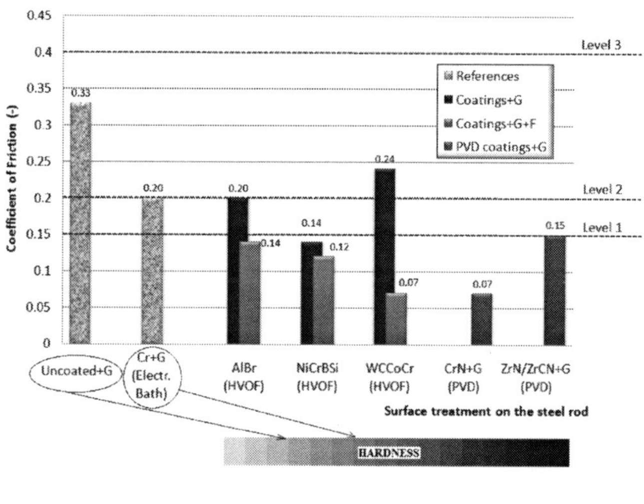

Figure 2: Coefficients of friction (CoF).

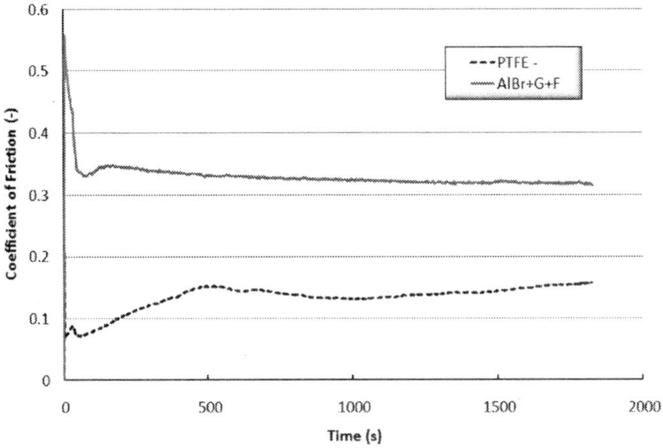

Figure 3: Comparison between PTFE/NBR against the CoFs of AlBr+G+F rods.

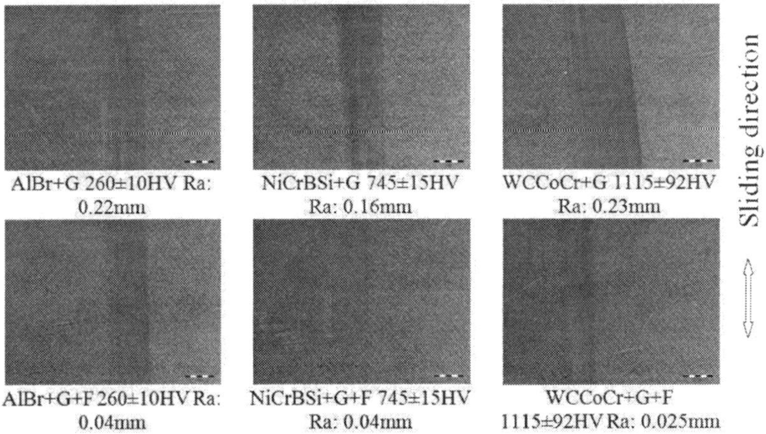

Figure 4: Wear of PTFE specimens.

Wear on PTFE specimens was observed and quantified: qualitatively, very similar wear mechanisms were observed independently of countersurface materials, coating, texture and hardness as can be seen for the example in Figure 4. The PTFE material was worn homogenously and the ploughing effect of the counterbody provoked defined wear scars. It seemed it was not able to form a thick transfer film on the countersurface and melting of the material did not occur.

Adhesion of particles was observed only when sliding against the AlBr rods: these surface films result from points on the PTFE and countersurface being brought into adhesive contact with subsequent sliding (Figure 5).

Figure 5: Adhesion of particles.

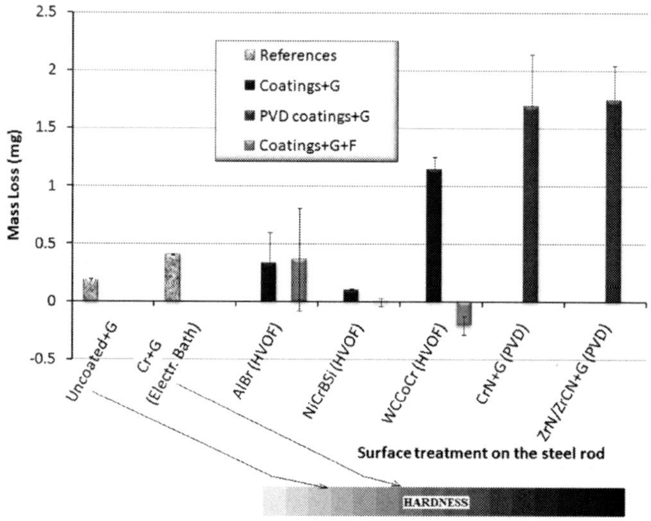

Figure 6: PTFE mass loss.

As can be seen in Figure 6, the higher wear on the PTFE samples is obtained when sliding against the hardest counterbodies.

On increasing the sliding displacement, PTFE fibrils broke along their length, leaving a fragment running along the surface, as well as being transferred to the countersurface, oriented along the sliding direction. Such fragments fractionally cover the countersurface filling in the roughness valleys to form a transfer film; the phenomenon, deeply explained in [5] for natural rubber sliding against metals, came out also for the PTFE material attacking the most reactive coatings (AlBr), independently of the surface finishing process applied. In fact, a transference film was found and consequently a chemical reaction film was formed on the steel surface. The other rods suffered just light surface polishing, also due to their higher hardness, as seen for example in Figure 7.

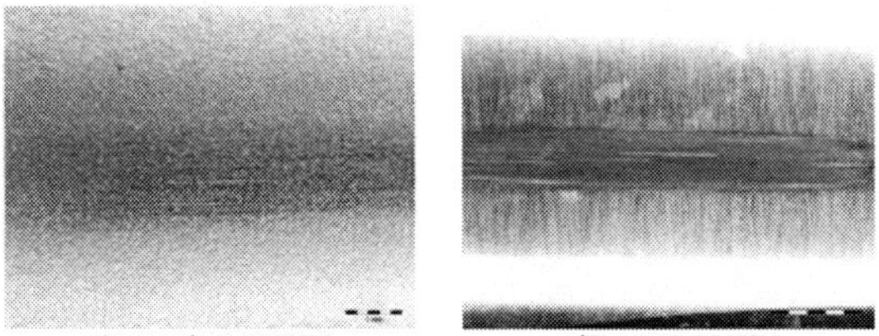

Figure 7: Wear of the rods' surface by PTFE.

4 Conclusions

A correlation of tribological measurements with surface characteristics, in particular coating material, roughness and hardness of a landing gear rod sealing system, was carried out. The finishing treatment effect on the CoF was observed to be the opposite of that met previously for NBR material, independently of rod hardness, as a demonstration that NBR has less ability to form a transfer film on the counterface, resulting in friction force increment. A relevant difference between NBR rubber and PTFE was observed in terms of the CoF at the initial stage of the test when the high degree of lateral mobility of PTFE chains quickly induces molecular orientation in the sliding direction.

Very similar wear mechanisms were observed independently of countersurface materials, coating, texture and hardness: the PTFE material was worn homogenously and the ploughing effect of the counterbody provoked defined wear scars. Just some spare particles were observed on the countersurface and melting of the material did not occur. The adhesion of particles was observed only when sliding against the AlBr rods: these surface films result from points on the PTFE and countersurface being brought into adhesive contact with the subsequent sliding.

Acknowledgements

The authors are grateful to all the partners of the KRISTAL (*Knowledge-based Radical Innovation Surfacing for Tribology and Advanced Lubrication*) and FUNDTRIBO (*Fundamentals of Tribology – Correlation between Wear Characteristics and Material Properties of Polymers Ranged from Sub-Micro to Macro Scale*) projects and the European Commission for their support.

References

[1] Calvert, C., Tirovic, M., Stolarski, T., Design and development of an elastomer-based pneumatic seal using finite element analysis. *Journal of Engineering Tribology,* **216(J3)**, pp. 127-138, 2002

[2] Belforte, G., Manuello, A., Mazza, L., Optimization of the Cross Section of an Elastomeric Seal for Pneumatic Cylinders. *Journal of Tribology,* **128**, pp. 406-413, 2006
[3] Conte, M., Interaction between seals and counterparts in pneumatic and hydraulic components. *PhD Thesis,* Turin, Italy, June 4th, 2009
[4] Fernandez-Diaz, B., Igartua, A., Fernandez, X., Mendoza, G., Straub, A., Tribological study of alternative CrVI-free hard coating sliding against polymers. *Proceeding of the World Tribology Congress 2009,* Kyoto, Japan, 2009
[5] Gent, A. N., Pultorf, C. T. R., Wear of metal by rubber. *Journal of Material Science,* **14**, pp. 1301-1307, 1979

Advances in aeropropulsion tribology and design using five key parameters

L. Wedeven[1], H. Chin[2], W. Ogden[2] & D. Haluck[3]
[1]*Wedeven Associates, Inc., USA*
[2]*Pratt & Whitney, USA*
[3]*Pratt & Whitney Rocketdyne, USA*

Abstract

Difficult tribology problems with wear, scuffing and contact fatigue in aeropropulsion are being addressed with a systematic approach. Solutions to these problems have been found using a systematic tribology approach involving detailed failure analysis and simulation testing. The risks in advanced development have been substantially reduced by the inclusion of tribology attributes in mechanical system design. These attributes control lubrication and failure mechanisms (wear, scuffing and contact fatigue). The developmental process involves five key tribology parameters. Simulation of wear, scuffing and micro-pitting can be accomplished with the control of the five key tribology parameters: entraining velocity (U_e), sliding velocity (U_s), film thickness-to-surface roughness ratio (h/σ), contact stress (both global and asperity scale) and contact temperature. Specialized test machines, test specimens and test protocols that control these parameters are able to replicate the failure mechanisms experienced in service. This approach has been used to develop next generation jet engine oil formulations that are compatible with stainless steel bearing materials. Simulation testing involving oil formulations, along with bearing steel composition and heat treatments, shows what technologies have the greatest impact on performance and where future design and development efforts should be focused.

Keywords: bearing steels, adhesive wear, aeropropulsion bearings, tribology testing, systematic tribology.

WIT Transactions on Engineering Sciences, Vol 66, © 2010 WIT Press
www.witpress.com, ISSN 1743-3533 (on-line)
doi:10.2495/TD100081

1 Introduction

Advances in aeropropulsion systems, such as those shown in Figure 1, are critically dependent upon a strong linkage between tribology and design. High power density and demand for reliability and long life require no tribology attribute to be left behind. A systems approach has evolved into a comprehensive process for staged technology readiness levels (TRL). The approach assures performance readiness before commitment to a new technology. It also creates an environment for innovation and a pathway to move good ideas toward maturity.

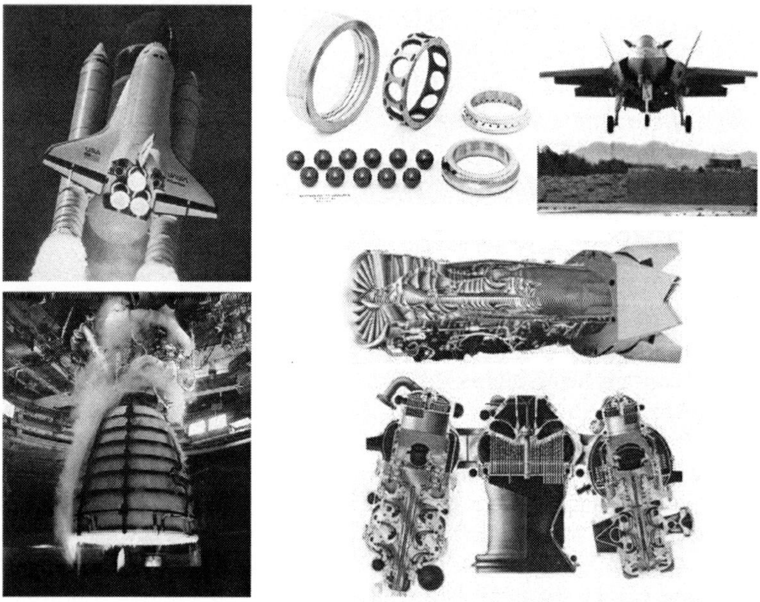

Figure 1: Aeropropulsion systems with demand for high power density.

Design of high stress lubricated contacts requires the capture of lubrication mechanisms and the avoidance of failure mechanisms. For many aeropropulsion components the margin between effective lubrication mechanisms and the onset of wear, scuffing and fatigue can be small. This paper shows how five key tribology parameters are used to define a linkage between tribology mechanisms and component design. The application of these five parameters is illustrated with an adhesive wear test protocol.

2 Background

In the past, development of advanced bearing materials was primarily focused on material characteristics for rolling contact fatigue resistance. Currently, the challenge is to develop other surface durability attributes that complement fatigue resistance attributes.

One can categorize surface deterioration into three basic modes: wear, scuffing and fatigue. The term "adhesive wear" is commonly used when failed surfaces appear to have undergone plastic flow due to local "adhesion" at the interface. The introduction of some advanced bearing materials, particularly corrosion resistant materials, has highlighted the importance of adhesive wear. Local adhesion and smearing events have been found to occur when the elastohydrodynamic (EHD) oil film thickness (h) is small relative to surface roughness height (σ). An example of adhesive wear near the shoulder of a thrust loaded ball bearing is shown in Figure 2.

If the surface chemistry of the material does not allow the formation of surface films from reactions with the oil, adhesive wear can supersede surface deterioration due to surface initiated fatigue. In addition, material properties that affect plastic flow, like hardness and the near-surface micro-structure seem to influence the onset of adhesive wear. The mechanisms that control the ability of a surface to handle high normal and tangential stress and to recover subsequent to local damaging events are not totally understood. Testing for these mechanisms and associated surface durability attributes is essential for material and oil formulation development to assure performance in service.

It was found that surface failure by adhesive wear is initiated at microscopic sites with insufficient surface film lubrication or at sites of debris encounters. With limited chemical reactivity between lubricating oil and some corrosion resistant materials, local adhesion events, which are not able to recover, propagate into broad patches of adhesive wear damage. With sufficient sliding velocity and contact stress, adhesive wear can transition into a major scuffing event. A scuffing event is characterized by a rapid rise in friction and temperature. These tribological features, as measured with an adhesive wear test method, correlate with experience in full-scale bearing tests.

At the heart of surface durability is material compatibility with lubricating oil chemistry to form surface films. These surface films are essential for preventing local adhesion. The adhesive wear test method described below progressively

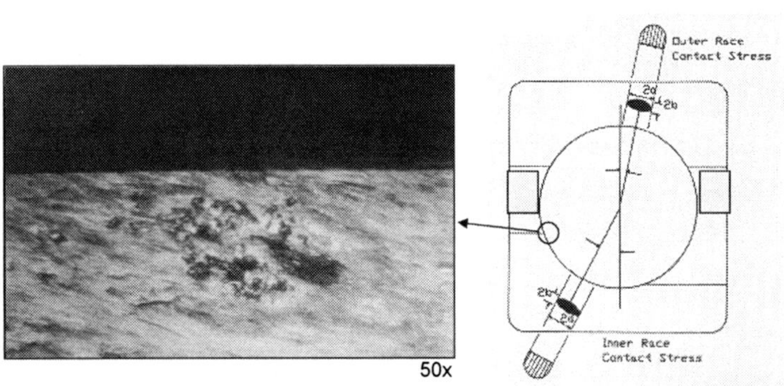

Figure 2: Adhesive wear – adhesion and plastic flow at the shoulder of a thrust loaded bearing.

increases the degree of asperity encounter at the interface under rolling/sliding conditions. The test method invokes tribological interactions, which are measured in terms of friction (traction), gentle polishing wear of surface features, adhesive wear events and scuffing. The test protocol described below is an attempt to simulate the adhesive wear mechanisms that are believed to occur in high speed rolling element bearings.

3 Systematic tribology

Historical approaches for bearing, gear and lubrication development have focused on design and development of "bulk" steel properties and "bulk" oil properties for durability. Today, development is directed toward the management of the tribology attributes of the contact interface. The contact interface is considered as a “tribo-system”, such that we can define four structural elements and technology regions, as shown in Figure 3.

These structural elements are: (1) hydrodynamic or elastohydrodynamic (EHD) film region; (2) surface film region; (3) near-surface region; and (4) subsurface region. These regions perform specific functions for lubrication and they provide the intrinsic properties for surface life and durability. For highly stressed lubricated contacts, durability is derived from three potentially active lubrication mechanisms which generate surface-separating films to carry the load and bear the shear at the interface. These lubrication mechanisms are: EHD, micro-EHD and surface film generation. Surface films, or boundary lubrication mechanisms, are required to preserve the integrity of contacting surfaces so that EHD and micro-EHD mechanisms can properly function. If boundary

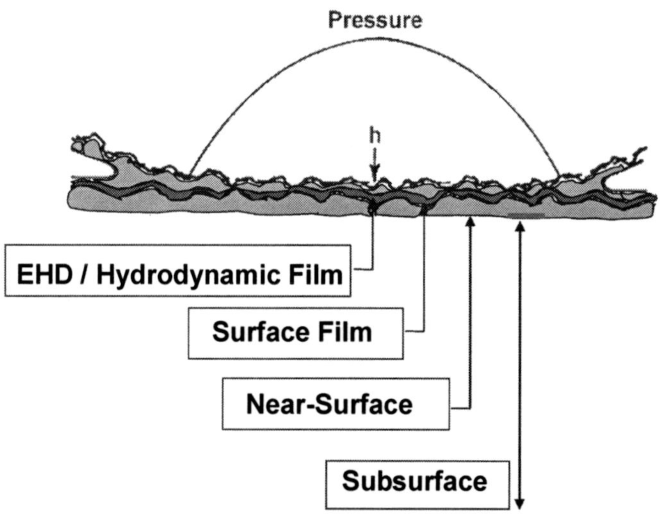

Figure 3: Structural elements of a contact.

lubrication mechanisms are compromised, the margin for lubrication of the total tribo-system is severely compromised. Experience has clearly shown that the durability of aeropropulsion hardware is critically dependent upon all three mechanisms. Since compatible oil and material chemistry is required to create effective surface films, oil attributes and bearing material attributes are fundamentally linked.

Durability is derived from the tolerance of stresses and strains to be accommodated at asperity sites. Durability is limited by multiple mechanisms reflecting performance attributes in terms of general categories of wear, scuffing and fatigue processes. Material or surface technologies to improve one attribute are frequently at the expense of another attribute. This has been experienced in both the gearing and bearing technical communities. It is important that no attribute is left behind. To assure wear, scuffing and fatigue performance, the structural elements must be tested as a system with linkage to design through parameters that the designer can use.

4 Five key parameters

The challenges of bearing and gear design are to invoke lubrication mechanisms of full-film EHD, partial-film EHD and boundary film lubrication and to avoid failure mechanisms associated with wear, scuffing and fatigue. Tribology testing must capture the micro-scale lubrication and failure mechanisms controlling performance and somehow link the testing results to the macro-scale design space that the design engineer can utilize. Five key tribology parameters illustrated in Figure 4 have been found to be effective in linking tribology mechanisms in service with testing and design. The five parameters are derived from EHD theory.

The entraining velocity (U_e) controls the formation of EHD film thickness through the generation of pressure in the inlet region. The sliding velocity (U_s) controls the shear and heat generation with the Hertzian region. The film thickness-to-surface roughness ratio (h/σ), or Lambda, controls the degree of

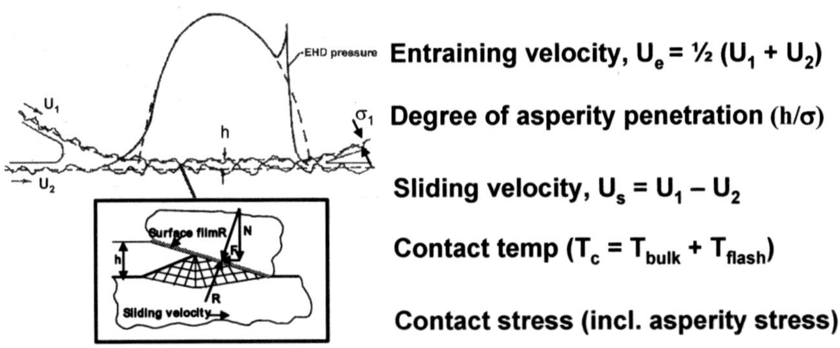

Figure 4: Five key parameters that link tribology mechanisms with design and testing.

asperity interaction. The contact stress is considered on a global Hertzian contact scale and a local asperity scale. Adhesive wear, micro-pitting and micro-scuffing are generally initiated on a local scale and subsequently propagated to a global scale. Contact temperature (T_c) is considered as the sum of the bulk temperature (T_b) and the friction generated flash temperature (T_f). The contact temperature controls reaction rate between the lubricant and surface material. It also controls the shear properties at the interface.

The five key parameters control both lubrication and failure mechanisms. They are useful for tribology design, failure analysis and testing. The ability to invoke these mechanisms in a simulation testing environment is accomplished with a patented [1] Wedeven Associates Machine (WAM) and test methodologies. The test machine is illustrated in Figure 5.

The test machine uses ball and disc test specimens with carefully prepared heat treatment, geometry, surface roughness and texture to simulate bearing or gear surfaces. Traction (friction) coefficient is measured along with bulk temperatures of the ball and disc specimens. The unique feature of WAM technology is the ability to decouple the entraining velocity, U_e from the sliding velocity, U_s. This allows precise control of EHD film thickness, the degree of asperity interaction (h/σ) and the shear rate within the Hertzian region.

Figure 6 shows a test plot with high nitrogen stainless steel specimens. The test is operated at low h/σ with 10.16 m/s entraining velocity (U_e), 2.48 GPa stress, 200 C and 15% slip (U_s = 1.52 m/s). The rise in traction coefficient with decreasing U_e is attributed to local surface damage caused by metallic debris roll-over. This particular high nitrogen steel is vulnerable to adhesive wear. The

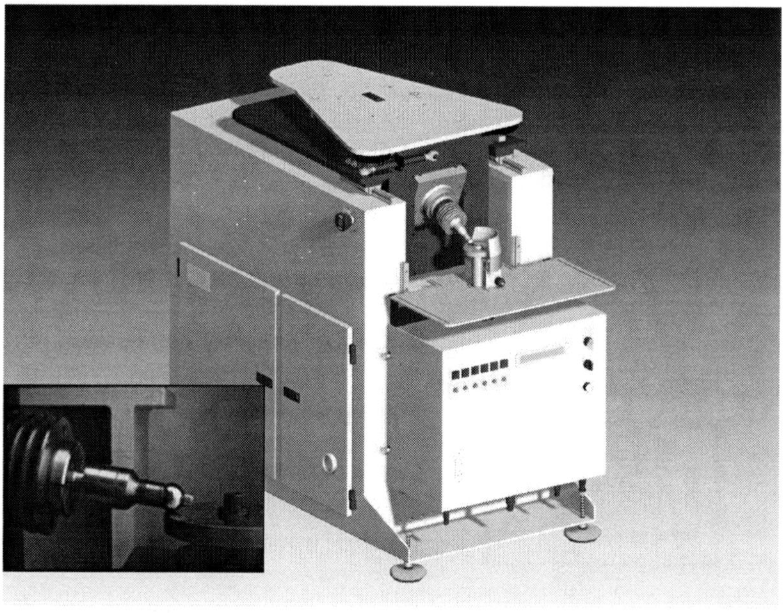

Figure 5: WAM test machine configuration.

test was terminated before the adhesive wear damage propagated to a more massive scale (see photomicrograph in Figure 6). A test protocol was developed to screen advanced bearing steels and lubricants for adhesive wear. The test protocol is shown in Figure 7 for a ball/disc material pair of M50/M50.

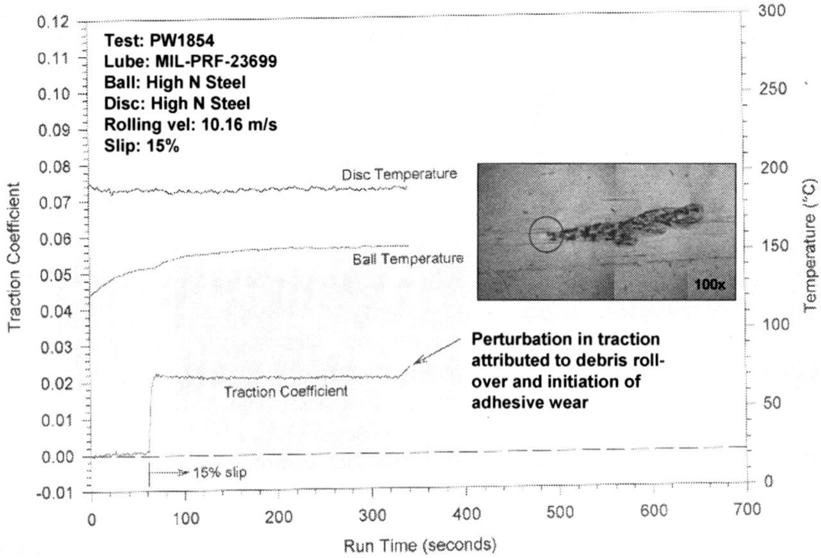

Figure 6: WAM test plot showing adhesive wear initiation at a local debris roll-over site.

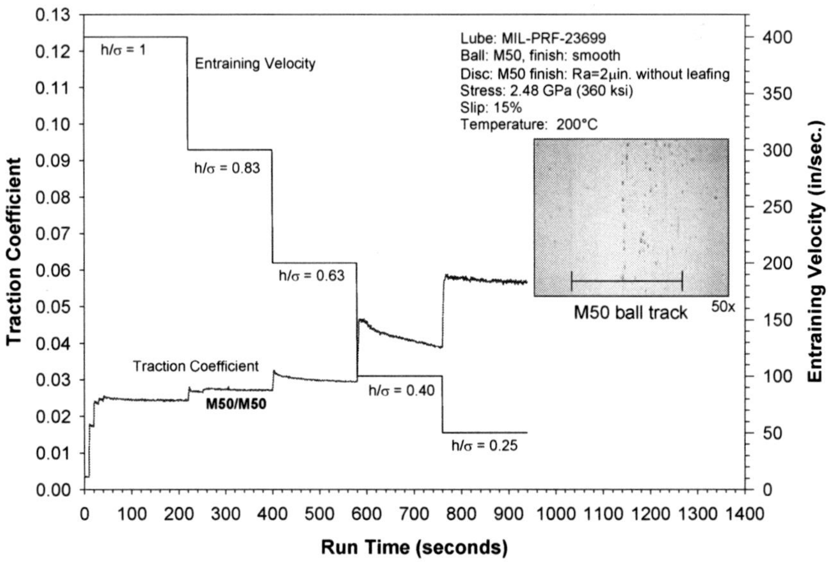

Figure 7: Test protocol for adhesive wear.

The disc surface finish is precisely honed to a roughness of 0.05 μm, Ra. The ball has an as-received lapped finish of < 0.025 μm, Ra. The test protocol reduces h/σ in progressive steps from an initial value of 1.0 to a value of 0.25. The test is run with a constant value of percent slip (15%). The test is conducted at 200 C and a maximum Hertzian stress of 2.48 GPa. At h/σ of 1.0 the traction coefficient is 0.025. This is representative of the limiting shear strength of a full EHD film. As U_e is reduced, the traction coefficient increases due to greater asperity interaction. A decreasing traction coefficient, like that shown for h/σ of 0.40, is due to minor polishing of the roughness features on the disc specimen. The adhesive wear test protocol with M50/M50 does not show any remarkable surface damage (see photomicrograph of ball specimen in Figure 7).

A repeat test using a high nitrogen stainless steel (version 1) is shown in Figure 8, along with M50. Shortly after initial operation at h/σ of 1.0, the traction coefficient rises to approximately 0.07 due to an adhesive wear event. The local adhesion and plastic flow disturbs the integrity of the surface. While the traction coefficient shows some recovery, the surfaces and traction coefficient do not recover to their original state (see photomicrograph in Figure 8).

For those materials which pass the adhesive wear test at 15 percent slip the test is repeated at 30 percent slip and subsequently at 50% slip. To avoid oxidation of the disc specimen at 200 C, the disc specimen is reconditioned with a light hone. The ball specimen is removed, documented and retested on the same track. The adhesive wear performance of material pairs and lubricants is judged by traction behaviour and surface deterioration features according to the criteria shown in Figure 9.

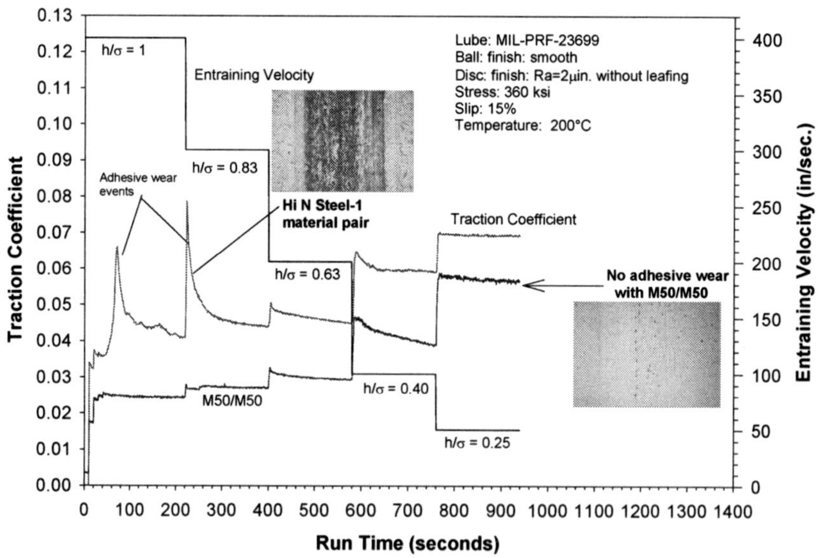

Figure 8: Adhesive wear test plot with high nitrogen steel (version 1) and M50/M50.

Measured Tribology Attribute	Weighting Factor
No adhesive wear events at 15% slip.	10
No large step increases in traction @ low λ (15% slip).	6
No rapid decline in traction coefficient reflecting advanced polishing wear.	4
Survives 8% slip without adhesive wear.	6
Survives 8% slip without major increase in traction.	3
Survives high slip (>15%) without adhesive wear events: 30% slip = 8; 50 % slip = 17	x
"Steady" traction behavior (no fluctuations due to a struggling contact).	2
Survives without a catastrophic scuff.	2

Figure 9: Adhesive wear performance criteria.

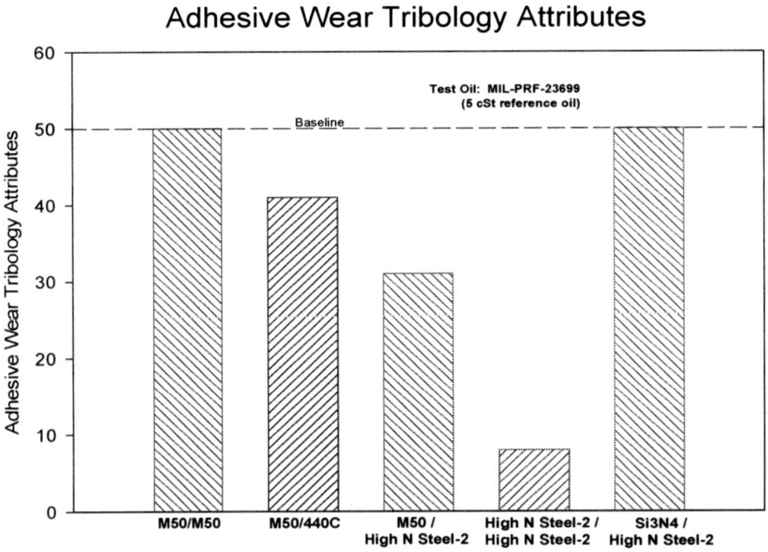

Figure 10: Adhesive wear for bearing material pairs.

Using the adhesive wear criteria in Figure 9, the test results for a combination of bearing materials are shown in Figure 10.

The mixed material pair of M50/440C shows reduced performance compared with M50/M50. The introduction of a stainless 440C material in the pair reduces adhesive wear performance. Adhesive wear performance is reduced further with high nitrogen steel (version 2). The introduction of silicon nitride with high nitrogen steel substantially improves adhesive wear.

At least some of the sensitivity of stainless steels to adhesive wear is the inherent passivation that chrome oxides have on the surface. The dominance of chrome oxides in place of iron oxides for non-corrosion resistant steels makes it difficult for anti-wear additives to sufficiently react with the steel to form protective surface films. To reduce this difficulty the bearing surfaces can be

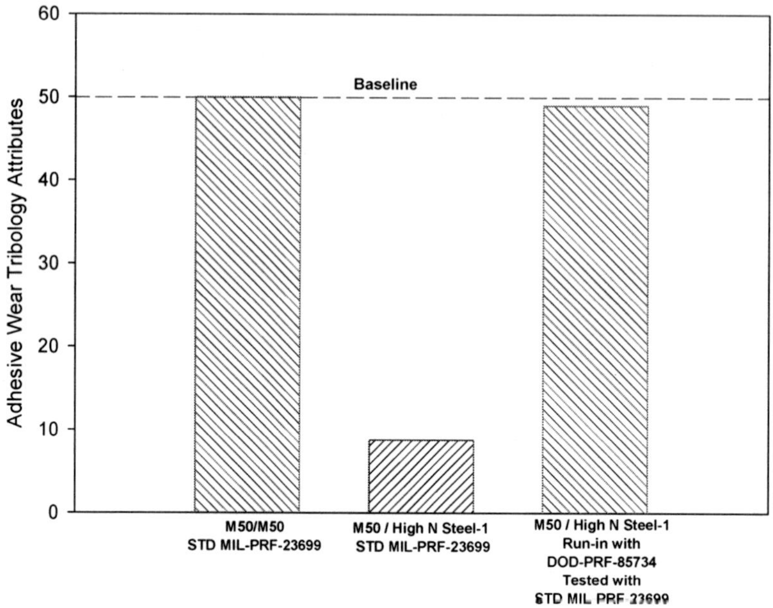

Figure 11: Effect of pre-test run-in on adhesive wear.

run-in with more aggressive additive formulations. This is shown in Figure 11. The data in Figure 11 shows that a good run-in with a more chemically active oil formulation (DOD-PRF-85734) is able to improve adhesive wear performance. This approach is limited as the run-in film is only effective so long as it is not worn off.

5 Discussion

The five key tribology parameters and the decoupling of U_e and U_s in testing provide a useful simulation testing approach for development, design and assurance of performance in service. An approach which simulates tribology mechanisms operative in service is not always intuitive. Engineers feel more comfortable with tests that operate under surface speeds and contact conditions they are familiar with. The decoupling of U_e and U_s provides a highly precise control of h/σ and the shear within the contact. This allows testing methods to cover all the major lubrication and failure modes. Low h/σ and high U_s are required to encounter scuffing failures. Lower values of U_s will introduce wear and micro-scuffing. The control of asperity stress and operation at low h/σ is a way to introduce micro-pitting. The adhesive wear test protocol is only one of several test protocols used for development and qualification testing. The importance of adhesive wear testing, along with testing for other tribology

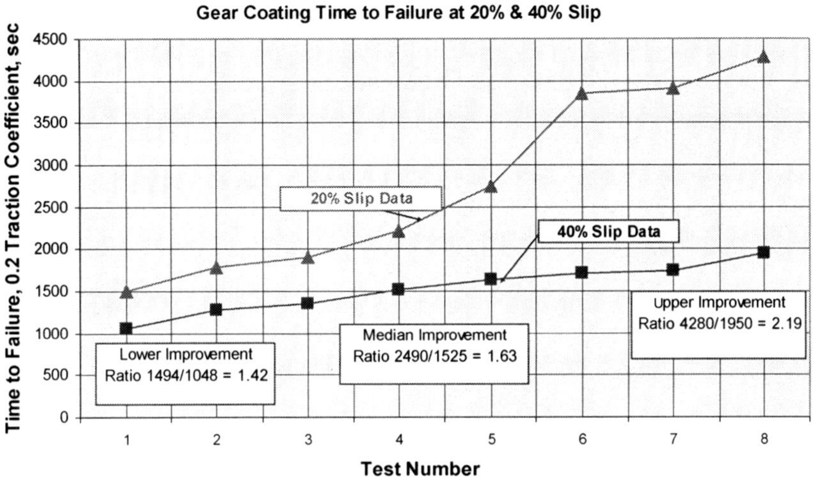

Figure 12: Cryogenic tests for rocket engine gears.

attributes, is associated with the management of performance attributes in design. For example, the adhesive wear vulnerability of high nitrogen steels can be managed. Excellent corrosion resistance and contact fatigue resistance make these materials highly desirable. Comprehensive tribology testing can map out the design space where these materials are highly effective.

Exploring the impact of key tribology parameters on performance is an excellent way to direct technical development. One example of this is the discovery of the importance of sliding velocity (U_s) in gear tooth design for a specific rocket engine upgrade. Performance tests conducted at two different sliding velocities shown in Figure 12 provided critical information for redesign of the gear tooth profile that gave substantially improved performance margin.

Many aeropropulsion systems have unknown motions or operating conditions at the interface. If performance issues need to be addressed, the five key tribology parameters are used in tests to explore conditions which cause the same surface deterioration features experienced in service. The actual operating conditions are deduced from experimentation.

6 Conclusions

Five key tribology parameters: entraining velocity (U_e), sliding velocity (U_s), film thickness-to-surface roughness ratio (h/σ), contact stress and contact temperature, provide links between tribology mechanisms in service, simulation testing and design. Unique testing capability that decouples U_e and U_s allows precise control of high stress lubricated contact conditions to invoke all the major lubrication and failure modes. This systematic tribology approach clearly defines the adhesive wear attributes of bearing materials. Additional test methods for other attributes provide a design space for tribology performance.

Reference

[1] Wedeven, L.D., "Method and Apparatus for Comprehensive Evaluation of Tribological Materials, U.S. Patent No. 5,679,883, October 21, 1997.

A new identification method of the supply hole discharge coefficient of gas bearings

G. Belforte, F. Colombo, T. Raparelli, A. Trivella & V. Viktorov
Department of Mechanics, Politecnico di Torino, Italy

Abstract

The paper shows a new methodology to identify the orifice discharge coefficient of pneumatic externally pressurized bearings. A common method of identification is to measure the experimental air flow, the supply pressure and the local maximum of the pressure distribution in the bearing clearance near the supply hole. This method however involves some practical difficulties especially with small supply hole diameters due to the high pressure and velocity gradients near the supply holes. The proposed new method requires both theoretical and experimental approaches. It differs from the previous method because the experimental activity involves measuring only the pressure in correspondence of one specified point under the pad far from supply holes. The pressure distribution under the pad is calculated by solving the Reynolds equation. The pressure boundary condition near the supply holes and consequently the pressure distribution in the pad clearance depend on the supply hole discharge coefficient. By using the bisection method the discharge coefficient value is changed iteratively in order to find the convergence of the experimental pressure value with the numerical one. The discharge coefficient is obtained when the experimental pressure measured in the selected point corresponds with the calculated one. This new method is simpler to realize experimentally and it is less sensitive to the position of the measuring hole. Air pads with different supply hole diameters are tested. The pressure distribution under the pads is determined experimentally using a test bench at the purpose realized. The tests are carried out with different air clearances and supply pressures.

Keywords: pneumostatic air pad, discharge flow coefficient, gas lubrication, gas bearing.

WIT Transactions on Engineering Sciences, Vol 66, © 2010 WIT Press
www.witpress.com, ISSN 1743-3533 (on-line)
doi:10.2495/TD100091

NOMENCLATURE

b	Critical pressure ratio
c_d	Discharge coefficient of supply orifices
d	Supply orifice diameter
d_0	Insert diameter
k_T	Temperature coefficient $=\sqrt{T^0/T}$
l	Supply orifice length
h	Air clearance
q	Inlet mass flow rate per unit area
r	Radial coordinate
N	Number of supply orifices
P	Absolute pressure
P_a	Absolute ambient pressure
P_s	Absolute supply pressure
P_c	Absolute supply orifice downstream pressure
R^0	Gas constant = 287.6 m^2/s^2K
T^0	Absolute temperature in normal condition
T	Orifice upstream absolute temperature
Re	Reynolds number
S	Supply orifice cross section
ϑ	Circumferential coordinate
ϕ_e	Pad external diameter
ϕ_i	Pad internal diameter
μ	Air viscosity = 17.89 10^{-6} Pa s

1 Introduction

Gas bearings are used for those applications in which high precision and repeatability in positioning are required. The absence of stick and slip and the very low friction makes these components very appropriate in precision linear guides. Externally pressurized air bearings are used to support measuring and testing machines and their static and dynamic performance is the object of continuous improvement in the design process.

While the equations that describe in good approximation the air flow in the clearance under the pad are well known, a problem not yet completely solved is the definition of the discharge coefficient of the supply holes. The researchers have dealt with this problem both experimentally and by means of numerical methods. In Al-Bauder and Van Brussel [1] an experimental activity on a centrally fed circular air pad is shown and compared with simulated results obtained with a method, called "separation of variables". A common approach consists in measuring empirically the discharge coefficient that relates the theoretical mass flow rate to the experimental one with the following formula

$$G_{\exp} = c_d G_{th}, \tag{1}$$

where the theoretical mass flow rate is calculated considering an isentropic expansion. Some papers present a single number, e.g. Lund [2] and Bryant et al. [3], while others propose empirical formulas, e.g. Elrod and Glanfield [4] and Kazimierski and Trojnarski [5]. Other authors like Renn and Hsiao [6], Belforte et al. [7] and Ly and Ding [8] solve the complete flow problem by using commercial CFD codes.

Belforte et al. [9] proposed a formula to calculate the discharge coefficient on the base of experimental measurements of mass flow rates and of pressure distributions near the supply holes. However this method presents some practical difficulties when the supply holes have a diameter smaller than 0.1 mm.

In the present paper is proposed another identification method that needs to measure experimentally the pressure only in correspondence of one point under the pad. This method is simpler to be realized and less time consuming because it doesn't need to measure the pressure distribution under the pad near the supply holes, where the pressure gradients are very high.

2 Pads under test

Four annular stainless steel pneumatic pads are examined. They have the same outer diameter of 40 mm and inner diameter of 10 mm. The central pocket is at ambient pressure.

Figure 1 shows the sketch of the pads and figure 2 their photo. In correspondence of the supply circumference of diameter 25 mm, N equispaced supply holes of diameter d and length l=0.4 mm are realized. The nominal diameter values are 0.2 mm, 0.3 mm, 0.4 mm and 0.2 mm for pads 1, 2, 3 and 4 respectively. The holes are drilled on brass inserts of diameter d_0 = 4 mm.

Table 1: Geometric parameters.

Pad type	Number of holes N	Orifice	Measured orifice diameter d_m [mm]	Mean diameter value [mm]	Roughness Ra [μm]
1	*3*	*1*	*0.237*	*0.222*	*0.45*
		2	*0.211*		
		3	*0.219*		
2	*3*	*1*	*0.312*	*0.313*	*0.46*
		2	*0.313*		
		3	*0.315*		
3	*3*	*1*	*0.421*	*0.420*	*0.61*
		2	*0.415*		
		3	*0.424*		
4	*6*	*1*	*0.228*	*0.221*	*0.38*
		2	*0.230*		
		3	*0.212*		
		4	*0.219*		
		5	*0.218*		
		6	*0.217*		

In Table 1 are shown the principal geometric parameters of the pads. Each supply hole was measured by a microscope along two perpendicular directions and the mean value d_m is reported in the table with also the measured mean diameter value of all the orifices. In the table are reported also the relative errors with respect to the nominal values and the roughness of the pads.

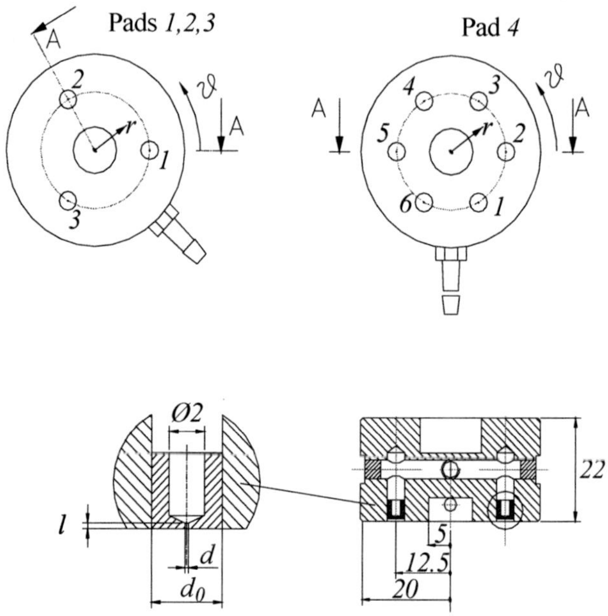

Figure 1: Sketch of the pads.

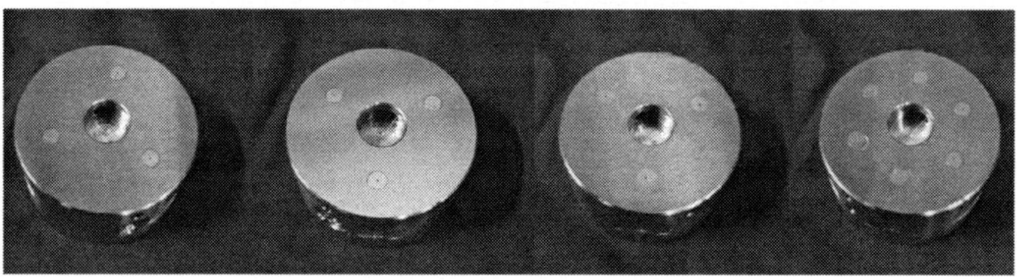

Figure 2: Photo of the pads.

3 Numerical model

The well known Reynolds equation (2) for compressible fluids is solved to calculate the pressure distribution under the pad.

$$\frac{1}{r}\frac{\partial}{\partial r}\left(rh^3\frac{\partial P^2}{\partial r}\right)+\frac{1}{r^2}\frac{\partial}{\partial \vartheta}\left(h^3\frac{\partial P^2}{\partial \vartheta}\right)+24\mu R^0T^0\frac{G}{rdrd\vartheta}=24\mu\frac{\partial(Ph)}{\partial t} \quad (2)$$

Mass flow rate G is calculated in accordance with the isentropic expansion equation (3), where P_c is the supply port downstream absolute pressure.

$$G = c_d G_{th} = c_d \pi \frac{d_s^2}{4} P_s \sqrt{\frac{k}{k-1}\left[\left(\frac{P_c}{P_s}\right)^{\frac{2}{k}} - \left(\frac{P_c}{P_s}\right)^{\frac{k+1}{k}}\right]\frac{2}{RT}} \text{ if } \frac{P_c}{P_s} \geq 0.528,$$

$$G = c_d G_{th} = c_d \pi \frac{d_s^2}{4} P_s \sqrt{\left(\frac{2}{k+1}\right)^{\frac{2}{k-1}}\left(\frac{k}{k+1}\right)\frac{2}{RT}} \text{ if } \frac{P_c}{P_s} < 0.528. \tag{3}$$

The discharge coefficient c_d can be identified experimentally. In paper [9] the following formula is proposed

$$c_d = 0.85\left(1 - e^{-8.2\frac{h}{d}}\right)\left(1 - e^{-0.001\,\mathrm{Re}}\right) \tag{4}$$

where Reynolds number is calculated with eqn (5)

$$\mathrm{Re} = \frac{4G}{\pi d \mu} \tag{5}$$

If formula (4) is used coefficient c_d is calculated on the base of the mass flow rate G of the previous iteration. Eqn (2) is discretized with finite difference method considering central derivatives. The surface under the pad is meshed with n and m nodes along the radial and the circumferential directions respectively. The position of the nodes is chosen in order to make some of them coincide with the supply holes. The Reynolds equation is solved with explicit Euler method with the initial condition of ambient pressure in correspondence of all nodes.

The following boundary conditions are considered:

- $P=P_a$ at the outer and inner edges of the pad
- coincidence of pressure for $\theta=0$ and $\theta=2\pi$.

The load carrying capacity of the pad is calculated integrating the relative pressure under the pad with eqn (6):

$$F = \int_0^{2\pi}\int_{r_i}^{r_e} (P - P_a)\, r\, dr\, d\theta \tag{6}$$

3.1 Discharge flow coefficient identification

In Belforte et al. [9] the discharge coefficient c_d was calculated with eqn (1) as the ratio between the experimental and the theoretical mass flow rate, obtained considering the experimental supply holes downstream pressure level P_c. This method however requires to measure the pressure distribution near the supply holes and individuate with good precision the downstream pressure level P_c. This could be difficult especially with small hole diameters because of the high pressure gradients. Moreover the measuring hole machined on the crossmember must have a smaller diameter than the supply hole in order to measure correctly the pressure distribution. This limits the diameter of the supply holes for the pads that can be tested correctly on the test bench.

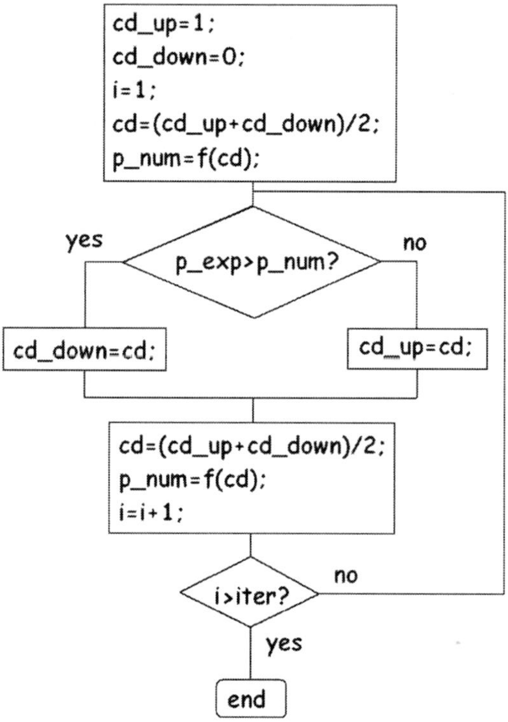

Figure 3: Block diagram of the identification algorithm.

A new method of identification of the supply holes discharge coefficient is proposed hereunder. The method consists in measuring the pressure P_{exp} in correspondence of a determined point under the pad for different clearance values. This pressure is compared with the numerical value P_{num} that one obtains with a discharge coefficient c_d of first approximation. The c_d value is then modified and the relative pressure is calculated numerically; different iterations are performed using the bisection method until convergence is reached. The final numerical discharge coefficient corresponds to a numerical pressure value equal to the experimental one. In figure 3 is shown the block diagram of the algorithm used. The number of iterations performed is *iter*=8, to which corresponds a convergence error $err<0.2\%$, where $err=(P_{exp}-P_{num})/P_{exp}$.

4 Experimental set-up

4.1 Test rig description

The test rig is composed of a rigid structure and a moving screw with which it is possible to impose different air clearance heights between the pad and the stationary member. The air gap is measured by means of two micrometric inductive probes with sensitivity 0.1 μm. The load carrying capacity of the pad is measured with a load cell with 2 kN full range. The pressure under the pad is measured with a piezoresistive sensor with resolution 10^{-4} MPa, connected to the measuring hole drilled in the stationary pad (see figure 4). The latter can be moved radially or circumferentially in order to measure the pressure value in

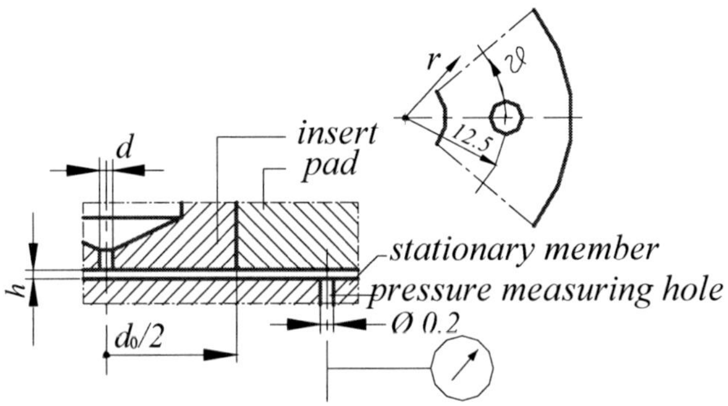

Figure 4: Sketch of the measuring system.

Figure 5: Test rig.

WIT Transactions on Engineering Sciences, Vol 66, © 2010 WIT Press
www.witpress.com, ISSN 1743-3533 (on-line)

each point under the pad. A potentiometric displacement transducer measures the radial position of the stationary member with respect to the center of the pad.

The air flow rate is measured by a flowmeter situated upstream the pad. High efficiency air filters (97% and 99% efficiency with particles of diameter 0.1 μm) are used to avoid problems of clogging of the supply holes. Figure 5 shows the test rig.

4.2 Test procedure

The load carrying capacity and the air consumption are strongly influenced by the air gap. For this reason a test procedure was defined in order to obtain a good repeatability and precision of the measurements.

The pad is initially cleaned and the supply holes are checked with a microscope to exclude an eventual clogging. The pad is then supplied with compressed air at a defined pressure and an increasing downwards vertical force is applied on it until the air clearance is equal to zero. In this condition the displacement transducers are set to zero. Higher loads are avoided in order not to deform the pad and the stationary member. In this way the reading of the displacement transducers coincides with the air clearance. The deformation of the pad and of the stationary member are negligible. The force applied to the pad is then decreased and the pressure under the pad, the load capacity and the air flow rate are measured for different air gaps. The supply pressure is maintained constant for each clearance value up to h=20 μm. The measuring hole is positioned far away the supply holes where the pressure gradients are low. In particular it was positioned in correspondence of the supply circumference in the middle of two supply holes.

4.3 Experimental results and discussion

The pressure in correspondence of the measuring point was measured at different clearance values with P_s=0.6 MPa. In figure 6 the measured pressures under the

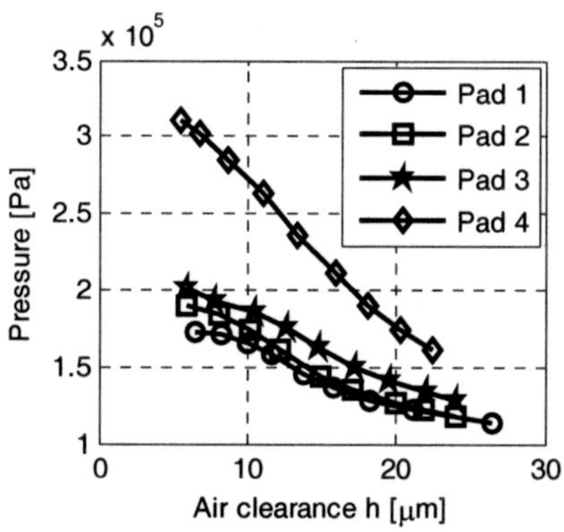

Figure 6: Experimental pressure in the measuring point under the pad vs air clearance; comparison between different pads.

four pads are compared vs the air clearance. As one can see, the pressure increases both with the diameter of supply holes (pads 1 to 3) and with the number of holes (pads 1 and 4). The second effect is predominant.

In figure 7 the air mass flow rate is reported vs the air clearance. The air consumption of pad 4 is almost twice than that of pad 1, being the number of holes twice and the supply hole diameter the same. In figure 8 the experimental load capacity of the pads are compared with the numerical ones. The numerical solution underestimates the experimental load capacity.

In figure 9 the values of the discharge coefficients obtained with the proposed method are reported. The values are compared with the ones obtained with the empirical formula (4). Considering that the value of c_d is very sensitive to the clearance value h, the comparison is quite good. The error between the formula and the experimental value is comparable with the tolerance admitted.

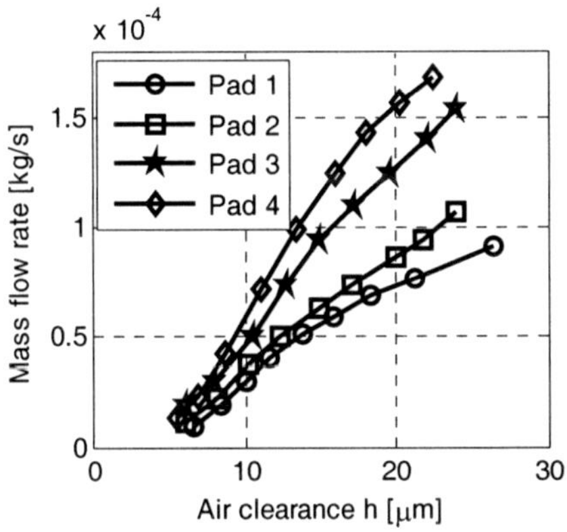

Figure 7: Experimental air consumption vs air clearance; comparison between different pads.

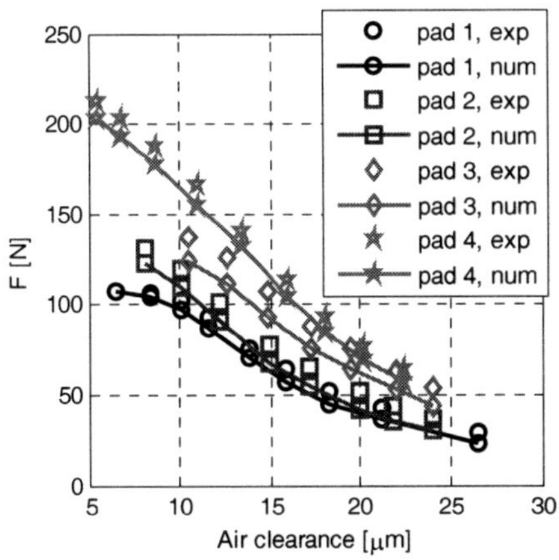

Figure 8: Load capacity vs air clearance; comparison between experimental and numerical results.

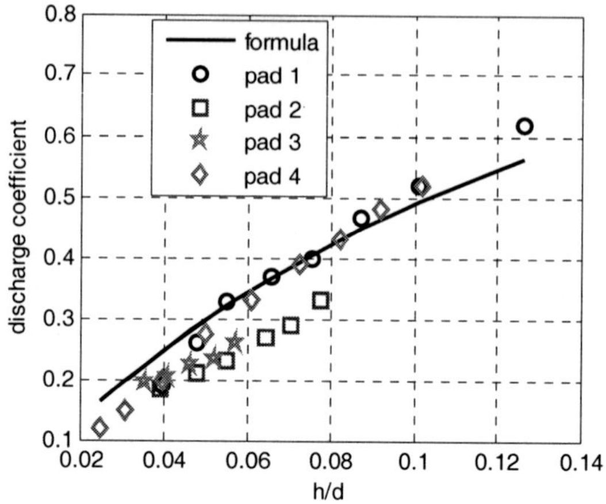

Figure 9: Discharge coefficient c_d vs air clearance; comparison between the values obtained with the identification method and with eqn (4).

5 Conclusions

A method of identification of the supply holes discharge coefficient for gas bearings is presented. The method allows the determination of the coefficient c_d without the need of measuring the complete pressure distribution under the supply holes. This involves some practical advantages that are more evident when the supply hole diameter is smaller than 0.1 mm. The diameter of the measuring hole is not needed to be smaller than the supply hole diameter because the pressure is measured far-off the supply holes, where the pressure gradients are not high. As the air consumption and the pressure under the pad are very sensitive to the air clearance, the experimental measures must be carried on with carefulness in order to have a good estimation of the discharge coefficient.

References

[1] Al-Bauder, F., Van Brussel H., *Symmetric radial laminar channel flow with particular reference to aerostatic bearings*, Journal of Tribology, **114**, pp. 630-636, 1992.

[2] Lund, J.W., *The hydrostatic gas journal bearing with journal rotation and vibration*, Journal of Basic Engineering, **86**, pp. 328-336, 1964.

[3] Bryant, M.R., Velinsky, S.A., Beachley, N.H., Froncza, K.F.T., *A design methodology for obtaining infinite stiffness in an aerostatic thrust bearing*, Journal of Mechanisms, Transmissions and Automation in Design, **108**, pp. 448-456, 1986.

[4] Elrod, H.G., Glanfield, G.H., *Computer procedures for the design of flexibly mounted, externally pressurized, gas lubricated journal bearing,* Gas bearing Symposium, University of Southampton, pp. 22.1-22.37, 1971.

[5] Kazimierski, Z., Trojnarski, J., *Investigation of externally pressurized gas bearing with different feeding systems,* Journal of Lubrication Technology, **102**, pp. 59-64, 1980.

[6] Renn., J., Hsiao, C., *Experimental and CFD study on the mass flow-rate characteristic of gas through orifice-type restrictor in aerostatic bearings,* Tribology International, **37**, pp. 309-315, 2004.

[7] Belforte, G., Raparelli, T., Trivella, A., Viktorov, V., Visconte, C., *Numerical analysis on the supply hole discharge coefficient in aerostatic bearings*, International Conference on Tribology AITC-AIT 2006, 20-22 September 2006, Parma, Italy.

[8] Ly, Y., Ding, H., *Influences of the geometrical parameters of aerostatic thrust bearing with pocketed orifice-type restrictor on its performance,* Tribology International, **40**, pp. 1120-1126, 2007.

[9] Belforte, G., Raparelli, T., Viktorov, V., Trivella, A., *Discharge coefficients of orifice-type restrictor for aerostatic bearings,* Tribology International, **40**, pp. 512-521, 2008.

Test rig for rod seals contact pressure measurement

G. Belforte[1], M. Conte[2], L. Mazza[1], T. Raparelli[1] & C. Visconte[1]
[1]*Department of Mechanics, Politecnico di Torino, Italy*
[2]*Manufacturing Processes Department - Tribology Unit, Tekniker, Spain*

Abstract

The sealing efficiency and the extent of the friction force in an elastomeric seal – counterpart sliding contact depend highly on the contact characteristics at the interface. In particular, a deep knowledge of the interface contact pressure is needed.

The work herein described deals with the design of a new test bench for an indirect contact pressure measurement. In particular, the radial force exerted by a rod seal sliding on a sensor-equipped portion of a cylinder rod is measured; during the seal displacement, the contact area between the seal and the sensor-equipped rod increases, so that the radial force trend during time can be acquired. Taking the sliding velocity at a low constant value, contact pressure can be indirectly computed by differentiating the force signal. This way it is possible to perform measurements without introducing a third body between the seal and its counterpart, so avoiding any modification of the contact surface.

Initially, the test bench working principle is described, highlighting all critical aspects related to the radial force transmission from the seal to the load sensor; subsequently, the test bench requirements were defined. Finally, the design solution was presented.

Keywords: contact pressure, radial force measurements, seals, elastomers, pneumatics.

1 Introduction

Analyzing seals performances is a particularly complex and time consuming goal. In fact, there is a wide variability in terms of available materials, geometries and dimensions; furthermore, the system behaviour is highly

www.witpress.com, ISSN 1743-3533 (on-line)
doi:10.2495/TD100101

influenced by working conditions, that is by applied loads, acting pressure, greasing or lubrication, temperature and velocity. In particular, a deep knowledge of the contact characteristics at the rubber-metal interface is required, since both sealing efficiency and the extent of the friction force depend on contact pressure.

A number of numerical and experimental studies aimed at evaluating contact pressure between a seal and its counterpart were reported. Bignardi et al. [1] analytically computed contact pressure between a pneumatic lip seal and a cylinder rod using experimental data from photoelastic tests. Yang and Salant [2] developed an analytical procedure involving contact pressure calculation in order to evaluate the behaviour of a tandem reciprocating hydraulic rod seal. Misiewicz et al. [3] employed a piezoelectric pressure mapping system to measure contact area and contact pressure between an agriculture tyre and the soil, obtaining accurate distribution at the contact surface. Lee et al. [4] employed film sensors for measuring contact area and maximum contact pressure on a lip seal for hydraulic rotating shafts, under different interference fits. Pressure sensitive films were also employed for static measurements of local contact pressure in pneumatic applications by Conte [5] and by Belforte et al. [6, 7]; it was seen that these sensors were effective in firstly estimating the contact characteristics (contact pressure and contact area). Nevertheless, a limit of film sensors is due to their non-negligible thickness and to the difficulty of positioning them between the two parts in contact [8].

Some other authors reported non intrusive experimental methods. An ultrasonic technique was used by Marshall et al. [9] in a bolted connection, with the advantage of actual contact conditions determination. Poll et al. [10] developed a test bench for indirect contact pressure measurement. The radial force exerted by a seal mounted on a sensor-equipped rod was recorded over time and contact pressure was obtained by differentiating this signal; the rubber-like material viscoelastic influence was minimized, because there was nearly no additional deformation of the seal during the measurement. This radial force meter required a force sensor with a special design and no pressure load could be applied to the seal.

The work herein presented refers to the design and manufacturing of a new test bench for contact pressure measuring on a rod lip seal for pneumatic actuators (rod diameter = 20 mm). The test bench is based on the same working principle described in [10] but adopts different design solutions, with the aim of minimising as much as possible additional friction forces that would affect the measurements. In particular, cheaper load sensors were used; in fact, the test bench was equipped with two common uniaxial load cells instead of suitably designed sensors. Furthermore, the test bench configuration allows a pressure load to be applied on the seal during the measurement.

In the following, the test bench working principle was firstly described. Subsequently, all critical aspects related to radial force transmission from the seal to the load sensor were discussed and test bench requirements were defined. Finally, the test bench design solution was presented.

2 Measuring principle and test bench requirements

The experimental test consisted in acquiring over time the radial force that a lip seal exerts on a sensor-equipped portion of a cylinder rod, while a rod seal slid on it at constant velocity. The sensor-equipped rod can be obtained splitting the rod into two sectors, a sensor-equipped part and a fixed parte.

The contact area was scanned during the seal motion. Figure 1 shows the seal at the beginning of the test (a), during the measurement phase (b), i.e. when the seal is gradually engaging the sensor-equipped rod, and at the end of the test (c).

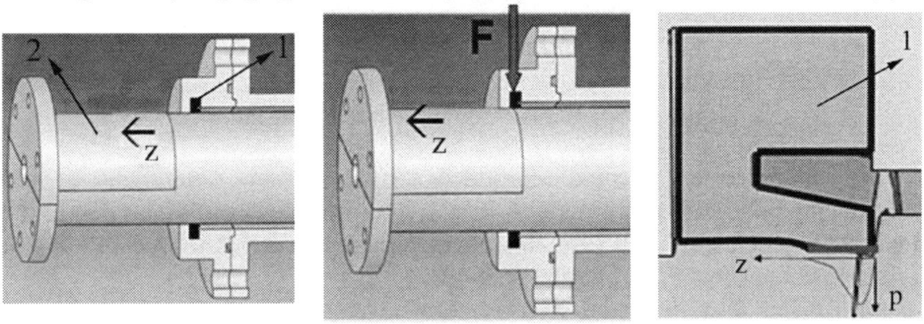

a) b) c)

Figure 1: Measuring sequence.

Since the seal (1) gradually slides in z direction on the sensor-equipped part of the rod (2), a gradual increase of the contact area occurs.

Consequently, an increasing of the radial force F can be detected, due to the exerted contact pressure, until a final force value is reached; this value corresponds to a complete contact of the lip seal, of length L, with the sensor-equipped rod. Thanks to the constant velocity motion, it is possible to correlate the time-radial force trend with the z direction-contact pressure p trend along the contact surface. In fact, contact pressure distribution between the seal and the counterpart can be worked out differentiating the acquired force signal. With reference to Figure 2, representing a portion of the sensor-equipped rod (diameter D), the measured force F can be worked out as the resultant of the contact pressure p distribution over the semi-rod circular surface.
The balance equation (1) along the y direction is:

$$dF(z) = dz \cdot 2 \cdot \int_0^{\frac{\pi}{2}} p(z) \cdot \frac{D}{2} \cdot \sin\alpha \cdot d\alpha \tag{1}$$

By integrating, it will be obtained:

$$dF(z) = dz \cdot p(z) \cdot D \tag{2}$$

Dividing both terms of (2) by time dt, p(x) can be obtained as:

$$p(z) = \frac{dF(z)}{dt} \cdot \frac{dt}{dx} \cdot \frac{1}{D} = \frac{dF(z)}{dt} \cdot \frac{1}{V \cdot D} \tag{3}$$

where V is the constant sliding velocity of the seal along the z-direction.

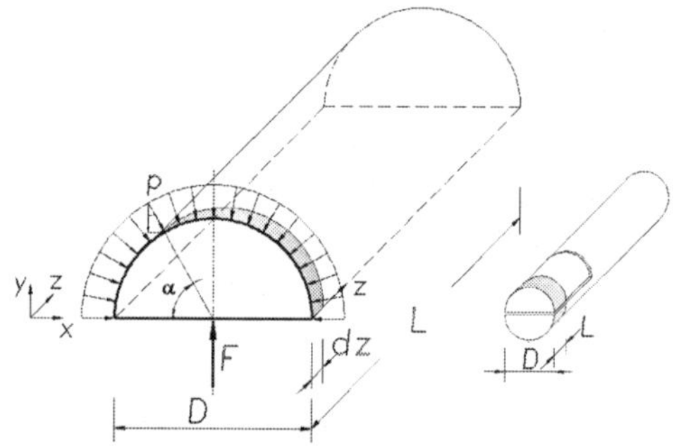

Figure 2: Sensor-equipped portion of the rod.

This measurement method reproduces actual working conditions without introducing additional deformations, so avoiding friction contribution due to the material viscoelastic behaviour; this characteristic represents a great advantage with respect to other experimental methods.

Measuring also the axial friction force exerted by the seal, the friction coefficient could also be determined.

In order to be able to measure the seal radial force as described previously, the cylinder rod must be cut in two portions, one of them connected to a load sensor. Figure 3 shows the chosen configuration, made up by: a fixed rod portion; a movable rod portion, integral with a plate for radial force transmission, denoted in the following as the "movable group".

A first requirement consisted of reproducing the rod nominal diameter, so to ensure nominal contact conditions between the seal and the rod. A gap h_1, with parallel surfaces A and B, was provided between the fixed and the movable rod portions in order to transmit the radial force, which tend to press the two portions one against the other. Values of h_1 in a range of 20÷50 μm can be considered sufficient for allowing radial force transmission to a load cell with a defined stiffness and for avoiding the seal intrusion inside the gap. Another gap of height h_2 was provided between surfaces C and D, in order to permit free (without friction) relative displacements. A pair of very low friction guide-bearings ensured guiding the movable group displacement along the seal radial force direction (y-direction). In Figure 3, the seal radial force and the seal friction force are denoted as Fy and Fz respectively.

The seal friction force Fz is balanced by the reaction forces R_A and R_B of the low friction guides. The seal radial force Fy is balanced by the reaction force R_L of a load cell placed between the movable group and the fixed frame (not shown in Figure 3). Since during the measurement Fy moves along the z-axis, Fy and R_L act as a couple, causing a moment to arise on the movable group; the low friction guides also balanced this moment.

A low friction pad avoided the mobile group rotation about the y-axis, thus ensuring the correct relative position of the two portions of the rod, i.e. the rod cylindrical shape.

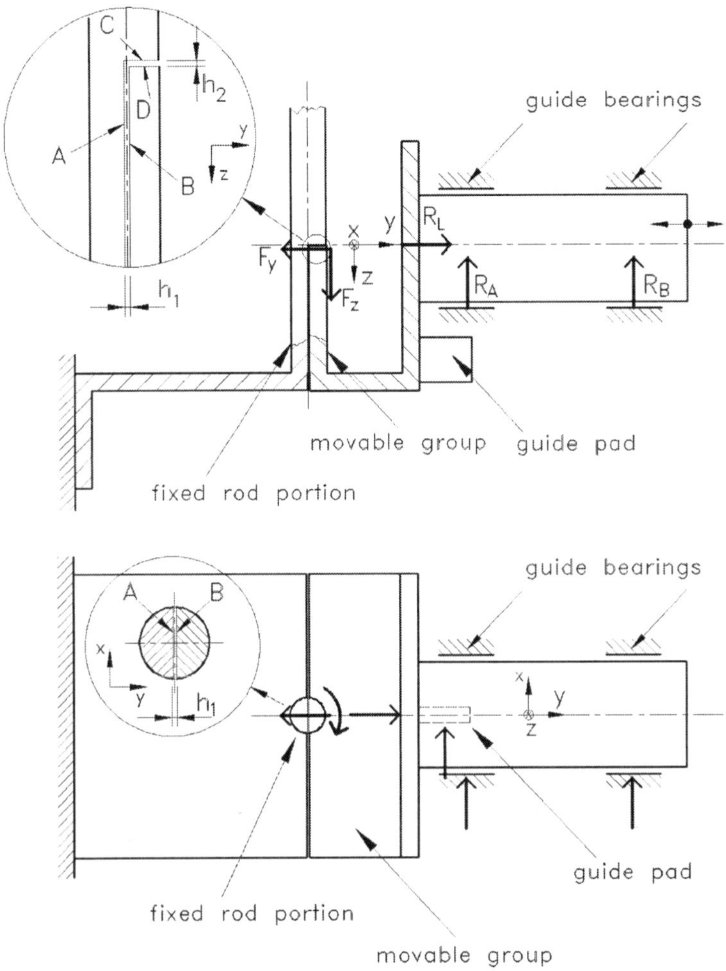

Figure 3: Test bench fuctional scheme: geometry requirements.

The load cell deformation must be less than the nominal gap h_1; to this aim, a load cell with a defined stiffness should be selected.

3 Test bench structure

A special test bench for an indirect measurement of contact pressure on a lip seal for pneumatic cylinder rod was designed and manufactured. It can reproduce actual working conditions. A vertical configuration was chosen, as shown in Figure 4. The bench can be divided into three main parts: the actuation part on the top, the measuring part at the bottom and the vertical fixed frame.

The seal motion is driven by a linear electric actuator, whose velocity can be controlled within a range of 0,5-300 mm/s. The actuator is equipped with an antirotation system and with a wire-type sensor, for displacement and velocity recording.

A pneumostatic bearing (1) allows guiding the seal holder (2) linear motion without additional friction forces, to permit a regular seal displacement even at very low velocities. Connection between the linear bearing (1) and the driving actuator is provided by a universal joint (3) and a spacing collar (4).

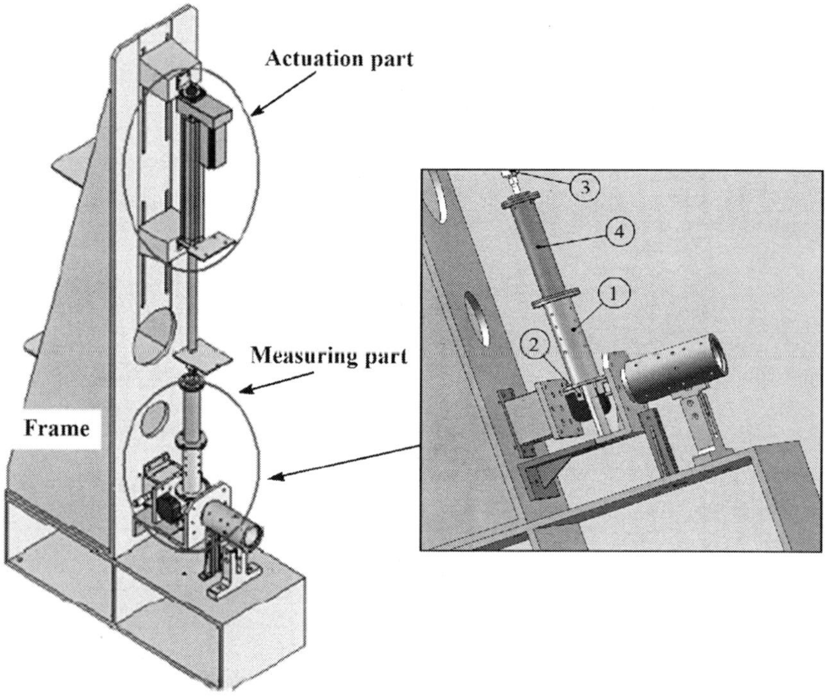

Figure 4: Test bench structure.

A detail of the test bench measuring part is shown in Figure 5. During the test, the seal is driven at first over the fixed part of the measuring rod (5) and then over the sensor-equipped rod portion (6). A couple of load cells (7), one of them shown in Figure 5, record the time-radial force trend during the seal motion. The movable group is supported by a pneumostatic bearing (8) and by a pneumostatic pad (9), with the aims described in the previous section. This way, a correct force transmission from the sensor-equipped rod to the load cells is ensured. Since a very precise relative positioning of the various components is required, a micro-positioner (10) and micro-metric screws (11) are used, together with calibrated metal shims. A nominal height of 50 μm was chosen for the gaps h_1 and h_2; this value was considered a good trade-off between the need of good precision measurements and acceptable machining costs.

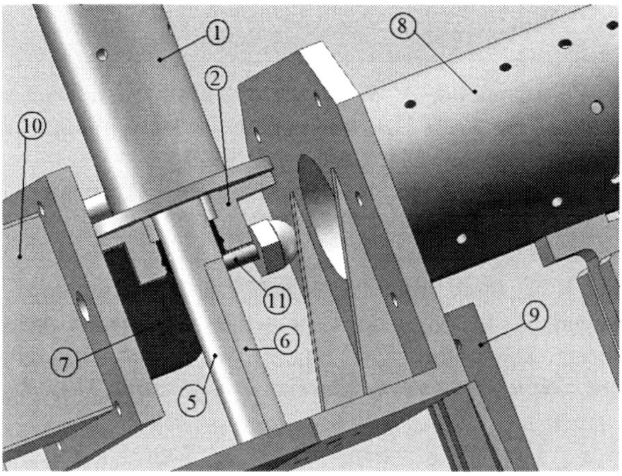

Figure 5: Details of the test bench measuring part.

The seal holder (2) is provided by a chamber that can be connected to the compressed air supply line, so that a pressure load can be applied on the seal. The chamber sealing is ensured on one side by the seal under test; on the opposite side, a small air leakage is allowed through a very narrow gap between the seal holder and the rod. Consequently, regulating the chamber supply pressure, it is possible to vary the seal pressure load.

4 Conclusions

A new test bench for an indirect contact pressure measurement between a pneumatic rod seal and its counterpart was designed. A rod seal is made sliding at slow constant velocity over a special rod, made up of a fixed portion and of a sensor-equipped portion; the latter allows recording during time, i.e. during the increase of contact area between the seal and the sensor-equipped rod, the radial force exerted by the seal on the rod. Differentiating in time this signal, it is possible to obtain indirectly the contact pressure distribution over the seal-rod contact surface.

The chosen measuring method has the great advantage of avoiding the interposition of sensors between the seal and the counterpart, so that the contact conditions are not modified; furthermore, a pressure load on the seal under test can be applied.

In order to perform the measurements correctly, any friction source different from the seal itself must be avoided. To this aim, pneumostatic components were provided and special design solutions were chosen.

A critical aspect is the test bench assembling: very narrow gaps must be provided between some parts and parts alignments must be done very carefully. To this aim, the test bench was provided by a precision linear positioner and micro-metric screws; calibrated metal shims were also used.

References

[1] Bignardi, C., Manuello, A. & Mazza, L., Photoelastic measurements and computation of the stress field and contact pressure in a pneumatic lip seal. *Tribology International,* **32(1)**, pp. 1–13, 1999.

[2] Yang, B. & Salant, R. F., Numerical model of a tandem reciprocating hydraulic rod seal. *Journal of Tribology*, **130(3)**, pp. 032201.1 – 032201.7, 2008.

[3] Misiewicz, P. A., Richards, T. E., Blackburn, K. , Brighton, J. L., Hann, M. J. & Godwin, R. J., Techniques for estimating contact pressure resulting from loaded agriculture tyres. *American Society of Agricultural and Biological Engineers Annual International Meeting 2008*, **1**, pp. 521-535, 2008.

[4] Lee, C. Y., Lin, C. S., Jian, R. Q. & Wen C. Y., Simulation and experimentation of the contact width and pressure distribution of lip seals. *Tribology International*, **39(9)**, pp. 915-920, 2006.

[5] Conte M., Interaction between seals and counterparts in pneumatic and hydraulic components. PhD Thesis, Politecnico di Torino, June 2009.

[6] Belforte, G., Conte, M., Manuello, A., Mazza, L. & Visconte, C., Experimental and numerical evaluation of contact pressure in pneumatic seals. *Tribology International*, **42(1)**, pp.169-175, 2009.

[7] Belforte, G., Conte, M. & Mazza L., Study on pneumatic cylinder piston seals behaviour. *15th International Sealing Conference*, Stuttgart, Germany, 7-8 October 2008.

[8] Wu, J. Z., Herzog, W. & Epstein M., Effect of inserting a pressensor film into articular joints on the actual contact mechanics. *Journal of Biomechanical Engineering*, **120**, pp.655-659, 1998.

[9] Marshall, M.B., Lewis, R. & Dwyer-Joyce, R. S., Characterisation of contact pressure distribution in bolted joints. *Strain*, **42(6)**, pp.31-43, 2006.

[10] Debler, C., Gronitzki, M. & Poll, G., Investigation into the sealing contacts of reciprocating elastomeric seals-correlation of calculations with measurements and optical observations. *17th International Conference on Fluid Sealing*, York, U.K., pp.169-186, 2003.

Section 3
Surface engineering

Evaluation of the friction properties of thermally sprayed coatings

M. Kašparová, F. Zahálka & Š. Houdková
Department of Thermal Spraying, ŠKODA VÝZKUM Ltd., Czech Republic

Abstract

In this paper, one of the tribological tests for the evaluation of friction properties is introduced. The test, named "Block-on-Ring", (in accordance with the standard ASTM G77) was carried out by the ŠKODA VÝZKUM Ltd. company and realized on renovated apparatus. The brief description of the measurement is as follows: the flat surface of the sample is pressed against the cylindrical surface of the rotating wheel. Arbitrary oil is fed to the materials' contact surface and the increasing temperature can be detected using the thermocouple that is placed in close proximity of the contact materials. Using the strain-gauge dynamometer, the friction force is recorded. The test is run on a number of periods and, including after the first expiration time, the wear track formed in the sample is measured and evaluated. The final results are the summary of partial results of the long-standing tests, which serve for the preliminary detection of the friction behaviour of the materials. The friction properties of the thermally sprayed coatings were investigated. The Cr_3C_2-NiCr, 13% Cr and Cr_2O_3 coatings were tested. The thin-base babbit was selected as the counter-part. The main interest was in the characterization of the friction properties in the dependence on the amount of added lubricant and on the normal load of sliding couples.
Keywords: friction, block-on-ring, thermally sprayed coatings.

1 Introduction

Friction properties rank among many important material properties which characterize the tribological behaviour of materials that are in a sliding contact. Because of the sliding between two or more machine parts friction arises. Friction is the resistance to motion during sliding or rolling that occurs when one

WIT Transactions on Engineering Sciences, Vol 66, © 2010 WIT Press
www.witpress.com, ISSN 1743-3533 (on-line)
doi:10.2495/TD100111

solid body moves tangentially over another one with which it is in contact. Then the resistive tangential force, which acts in a direction directly opposite to the direction of motion, is called the friction force. There are two main types of friction: dry friction and fluid friction. If the solid bodies are loaded together and a tangential force is applied, the value of the tangential force that is required to initiate the motion is the static friction force. It may take a few milliseconds before the relative motion is initiated at the interface. The tangential force required to maintain the relative motion is known as the kinetic or dynamic friction force. The static friction force is higher than the kinetic friction force, as Bhushan [1] showed. The science called "tribology" deals with main friction problems. A wide part of tribology science solves the following questions: surfaces interface, adhesive wear and lubrication of machine units. It is evident that in addition to the friction, wear also occurs. Vocel et al. [2] mentioned that friction without wear may occur if the hydrodynamic lubrication of function surfaces occurs and the compact lubricant layer is spread between two contact bodies. Friction and friction characteristics are advantageously measured using laboratory tribo-tests. The complete tests are described in the related Standards. The "Block-on-ring" test specified in the ASTM G77 [3] was used for determination of the friction properties of several selected thermally sprayed coatings (Cr_3C_2-NiCr cermet, Cr_2O_3 ceramic and 13%Cr stainless steel).

2 Experimental

2.1 "Block-on-ring" method

In ŠKODA VÝZKUM Ltd., the friction properties are evaluated using the "Block-on-Ring" apparatus, see fig.1b. The test is generally characterized in such a way that the flat surface of the sample (block) is pressed against the cylindrical surface of the rotating wheel of the rotors material (ring). Arbitrary oil is fed to the material contact surface and increasing temperature is detected using the thermocouple. The thermocouple is placed in the hole in the block sample. The hole is drilled 1 mm under the surface, in other words 1 mm under the samples contact, see Fig.2. The friction force is recorded using the strain-gauge dynamometer. A schematic draft of the test principle is depicted in Fig.1a. The test runs in a number of periods and always after one expiration time the wear track formed in the block-sample is measured and evaluated. One of the disadvantages of this apparatus is the constant velocity of the ring sample. Because of this, it is impossible to change the speed of the ring rotation and to record the statical coefficient of friction.

Using this apparatus it is possible to measure a dynamic friction coefficient in lubricated conditions, thermal changes in the materials contact, wear coefficient, the actual load of a sliding couple in the contact surfaces and it is possible to monitor the waveform of the friction force during the test running. The equations for determination these characteristic quantities are stated in [4, 9, 10].

For the experiment the nominal load of 1 MPa, 2,5 MPa and 3,5 MPa and sliding distance of 500 km was used. The total distance was divided in six cycles. The first cycle was the running-up step and it´s length was one kilometre. Each for further five steps was one-hundred kilometres long. The determination of the volume loss of the block sample is stated in ASTM G77 [3]. The scar width was measured by the magnifying glass with accuracy of measurement of 0,1 mm.

2.2 Experimental materials

Three samples of thermally sprayed coatings were chosen for the experimental measurement: one kind of cermet coating on the base of Cr_3C_2-carbide sprayed using "high-velocity-oxygen-fuel system" (HVOF), 13% Cr stainless steel sprayed using electric arc spray technology and ceramic coating of Cr_2O_3 sprayed using plasma spray technology.

The Cr_3C_2-NiCr coatings are commonly used as a protector against different types of wear up to approx. 850°C. Their superior oxidation resistance has seen them applied in a range of applications subject to combined erosion and high

a) b)

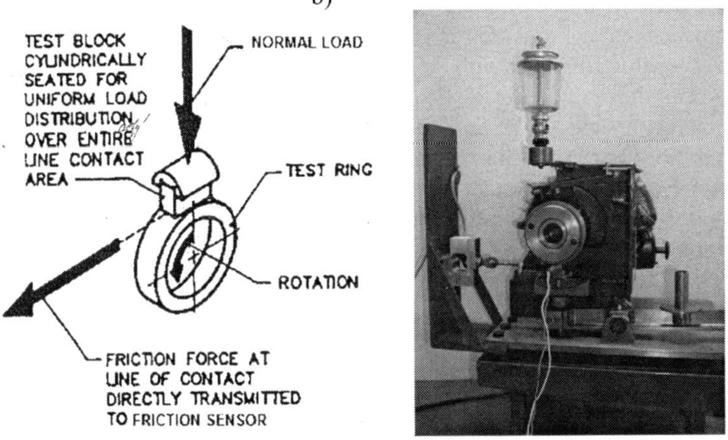

Figure 1: a) Test schematic – principle of the measuring of friction [3], b) "Block-on-ring" apparatus.

Figure 2: Temperature detection.

temperature corrosion [4], for example for turbine compression rings, jet engine components, replacing of hard chromium plating and so on. The coatings of 13%Cr stainless steel are sprayed on such surfaces where relatively good resistance of abrasion, corrosion and high-temperature oxidation is required. They are advantageously used for the pistons, headless bushings, bearing pins, etc. Plasma sprayed ceramic coatings have excellent performance in resisting high temperature, wear, corrosion and erosion. The Cr_2O_3 coating has good resistance in chemical environments and high resistance against the mechanical stress with low coefficient of friction. The ceramic coatings on the base of Cr_2O_3 are widely used as sealing parts of energetic, chemical and industrial machines. The surface hardness of tested coatings is mentioned in Table 1. The resultant values are the average values of three measurements.

The coatings were sprayed in ŠKODA VÝZKUM Ltd., Pilsen, CZ, only the plasma spraying was performed in Plasmametal, CZ. The used spray parameters are the intellectual property of the companies. The coatings were deposited on the test ring pieces of EN 10025-90 low-carbon steel.

The "stanit" tin-based babbit was chosen as the block sample. The "Stanit" tin-based babbit is one of many kinds of tin-compositions on the base of Sn-Sb-Cu. Its chemical composition is given in Table 2. The copper is very important additive for this alloy. Cu with Sn creates the chemical compound of Cu_6Sn_5 that crystallizes during cooling as first and it creates the dendritic skeleton. On the skeleton the crystals of the SnSb begin to grow [5]. Using the copper the crystals of SnSb are kept in the liquid by Cu_6Sn_5 that prevents to crystals segregation. Due to the copper the crystals of SnSb are after the crystallization homogenously distributed in whole structure. The SnSb together with Cu_6Sn_5 then creates the skeleton of the bearing [6–8].

2.3 Lubrication

The "Block-on-Ring" method facilitates to use many kinds of lubricants in a fluid stage. The lubricants, mostly oils, are injected into the surface contact of the samples. Oils include natural organics consisting of animal fat, vegetable oils, mineral fractions, synthetic organics and mixtures of two or more of these materials. It is mentioned in [1] that the main aim of the lubricants is the decreasing in friction losses, wear, noise level, and increase in absorbing and corrosive resistance. Currently in the industrial market many types of liquid

Table 1: Coating macro-hardness (HRC).

Cr_3C_2-NiCr	Cr_2O_3	13%Cr
61.8 ±2.3	68 ±3.1	32 ±1.8

Table 2: Chemical composition of the "stanit" tin-based babbit.

	Sb (%)	Cu (%)	Ni (%)	Pb (%)	Sn (%)
Stanit	9-11	2-4	0.4-1.0	max.1	84.5-86.5 + 0.5-1.1 Cd

lubricants are offered. From this large offering it is necessary to select the suitable lubricant for the practical application where moving parts with a sliding contact will work. The main tribological components in practical applications are for example, bearings, seals, cams, piston rings, electrical brushes, cutting and forming tools, etc. For the experiment the turbine deeply refined TSA 32 oil (in accordance with the ISO 6743 Standard) was chosen. This oil is characterized by the outstanding oxidative stability and antirust performance to protect turbines from rust and corrosion in severe conditions. It is suitable for steam turbines and water turbines, pumps and valves. The viscosity of the lubricant at temperature 40°C is 28,8 - 35,2 mm^2/sec. The thermal stability and its using are possible up to temperature 140°C. During the experiments the lubricant supply to the sliding contact was changed. The amount of lubricant was at all times sufficient for "the mixed friction." The test parameters are listed in Table 3.

2.4 Surface quality

Tribological characteristics depend on the surface quality of the sliding components, which are in relative contact (motion). Therefore, the surface roughness of the sliding parts is very distinctive factor. The "flat" surface shows lower opposition to the sliding then the "rough" surface and the friction force is naturally lower. Due to lower friction force the wear rate of the stanit is decreasing. If the comparing tests between several different materials are required it is preferable to compare sliding couples of the same surface roughness. The surface roughness was measured using SJ-201P Surftester and the Ra parameter was measured. The results are summarized in Table 4. The resultant values are the average values of five measurements.

3 Results of the measurements

The waveforms of the friction force and the temperature are the primary results of the friction test. By means of the Labview software the waveforms are displayed on the screen of the laboratory PC. The measurements values are saved to the Excel program simultaneously. Because of the values saving it is possible to calculate the average values of the friction force. The average friction force

Table 3: Parameters of measurements.

Test	1	2	3
Load [MPa]	1	2.5	3.5
Lubricant [dm^3/h]	0.018725	0.018725	0.002675

Table 4: Surface roughness of the samples.

Material	Cr_3C_2-NiCr	13%Cr	Cr_2O_3	Stanit
Ra [µm]	1.10E-01±1.55E-02	2.84E-01±2.07E-02	1.2E-01±1.05E-02	3.20E-01±5.00E-02

was calculated in each of the test step and the average values were recorded. The coefficient of friction was calculated and the results are recorded in the graphs, see Fig.3-5. In the following graphs the influence of the friction coefficient on the sliding distance is recorded. Figure 3 shows the dependence of the coefficient of friction on the sliding distance under the load of 1 MPa and amount of lubricant of 0,018725 dm^3/h and the Fig.4 under the load of 2.5 MPa and amount of lubricant of 0.018725 dm^3/h. Figure 5 shows the friction properties under the load of 3.5 MPa and of under reduced amount of lubricant of 0.002675 dm^3/h. For the reduced amount of lubricant the sliding distance was shortened up to 100 km to prevent the seizure.

It is evident that the coefficient of friction changes with the dependence on the type of coating. The fluctuation in the friction values conforms to the materials specific properties (hardness, roughness, porosity, phases

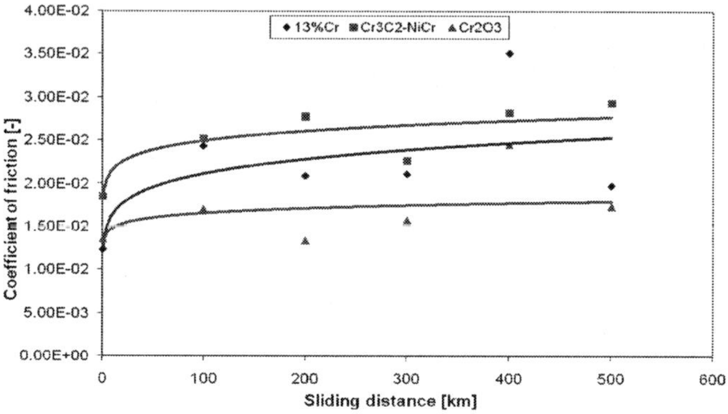

Figure 3: Dependence of the coefficient of friction on the sliding distance, load of 1 MPa, lubricant of 0.018725 dm^3/h.

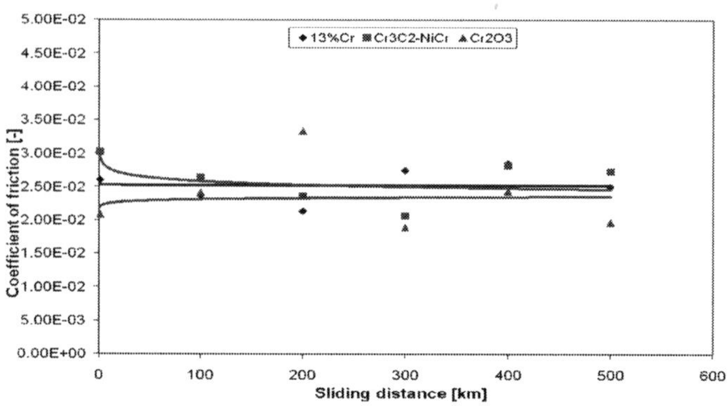

Figure 4: Dependence of the coefficient of friction on the sliding distance, load of 2.5 MPa, lubricant of 0.018725 dm^3/h.

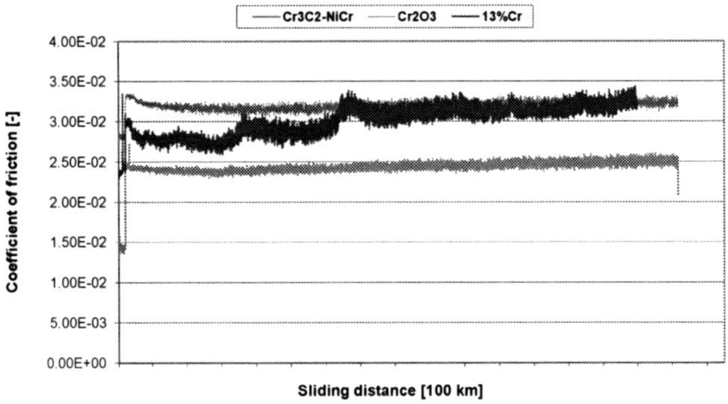

Figure 5: Dependence of the coefficient of friction on the sliding distance, load of 3.5 MPa, lubricant of 0.002675 dm^3/h.

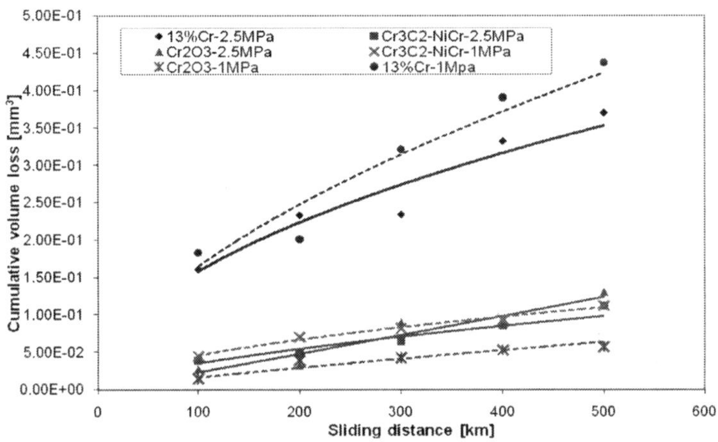

Figure 6: Wear of the block sample – stanit during sliding under the loads of 1 MPa and 2.5 MPa.

composition). The graphical dependencies show that the difference between the friction coefficients is not so significant. The difference between individual sliding couples is only about the value of 0.01. During sliding under the load of 1 MPa, Fig.3, the friction coefficient rises during the first 100 km of run and stabilises for all coatings after covering of 100 km. However, it is evident that increasing of the normal load leads to the stabilizing of the friction coefficient on the uniform value that is identical for all tested coatings (0.025), see Fig.4.

During sliding, the wear of the softer component of the sliding couple is generated. In our case the softer element is the block sample – stanit. The cumulative volume loss of the stanit is recorded in Fig. 6 and 7. In Fig. 6 the wear of the stanit under the load of 1 MPa and of 2.5 MPa and in Fig. 7 the wear

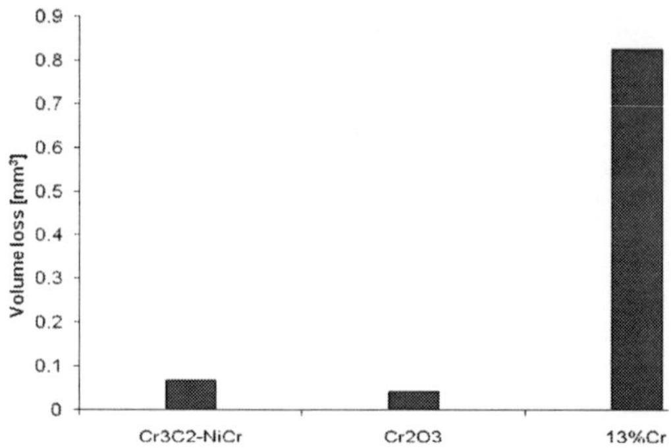

Figure 7: Wear of the block sample – stanit during sliding under the load of 3.5 MPa (100 km).

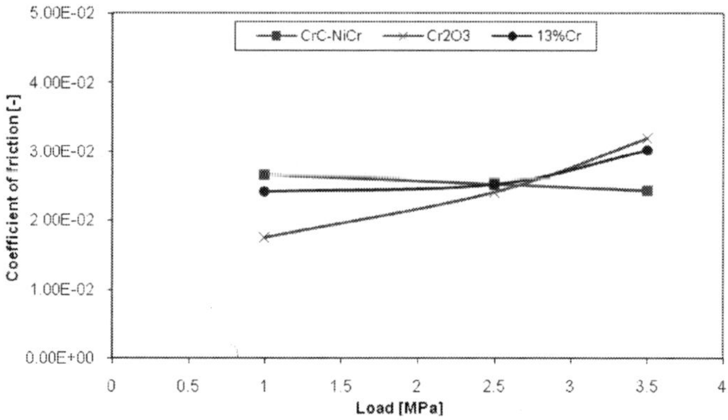

Figure 8: The dependence of the friction coefficient on the normal load.

of stanit under the load of 3.5 MPa is recorded. It is evident that the volume loss of the stanit blocks doesn′t change with the load change, Fig.6. The wear of the stanit uniformly increases in all tests. However, the coating of the 13% Cr stainless steel causes the most extensive wear of the stanit. In the test conditions of reduced supply of lubricant and higher load even eightfold rise of the stanits wear has been occurred, see Fig.7, whereas, the volume losses of the stanit which were caused by the Cr_2O_3 ceramic and Cr_3C_2-NiCr hardmetal were nearly identical.

4 Discussion

The coefficient of friction is given by the ration of the friction and the normal force. The coefficient of friction in the dependence on the load differs for individual sliding couples only in narrow range of values, see Fig.8. The slight values variation can be caused by the measurement error of the strain-gauge dynamometer. From the results it can be said that the coefficient of friction keeps constant with increasing load. Higher values of friction coefficient under the load of 3.5 MPa for Cr_2O_3 and 13%Cr coatings is probably due to lower lubricant supply.

Due to the friction and the load action it comes to the abrasive and adhesive wear of the softer part - stanit. The Fig. 9 shows the dependence of the stanit wear on the normal load. For the sliding couple of Cr_2O_3/stanit and Cr_3C_2-NiCr/stanit the wear of stanit-block increases with the increasing load linearly and the wear rise is very slight and uniform. The most aggressive counter-part for the stanit block is the coating of the 13%Cr stainless steel. This coating caused the highest wear of the stanit, especially in the test conditions of the load of 3.5 MPa and reduced lubricant supply. In the test conditions of the load of 1 MPa and 2.5 MPa (constant lubricant supply) the wear of the stanit kept constant. Identical trend was indicated for the Cr_3C_2-NiCr coating. Using C_3O_2 ceramic coating the constant stanit wear was recorded for the test conditions of the loads 2.5 MPa and 3.5 MPa.

Further factor that is necessary to monitor during sliding is the temperature that is accumulated in the sliding contact. Each of the sliding couples has their specific temperature that is influenced by the load, surface quality of the interacting materials, amount of lubricant and naturally also by materials properties of the sliding elements. In Fig. 10 the temperature in the dependence on the used normal load is recorded. The sliding couple of 13%Cr/stanit reaches the highest temperature during sliding and the rise of the temperature is even about 10 °C in the test conditions of reduced lubricant supply and increased load. It is evident that for coatings Cr_2O_3 and Cr_3C_2-NiCr the temperature doesn´t change so markedly. For these coatings the temperature slightly increases with increased load and reduced lubricant supply didn´t influenced the slight rising of the temperature.

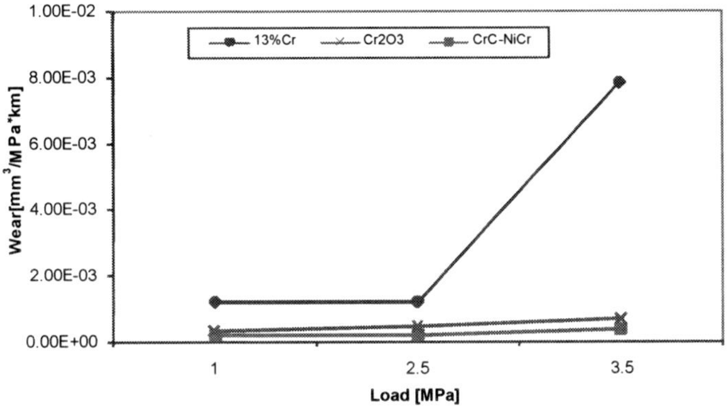

Figure 9: The dependence of the stanit wear on the normal load.

The friction properties of the HVOF sprayed coatings suitable for combustion engines, measured in compliance with ASTM G-99

Š. Houdková, M. Kašparová & F. Zahálka
ŠKODA VÝZKUM s.r.o., Czech Republic

Abstract

The friction between a piston, piston rings and a cylinder is responsible for almost 55% of total power loss in engines. The decrease of friction in the piston area leads to a significant fuel saving and also to emission reduction. Application of a surface treatment, such as PEO, PVD, HVOF, plasma sprayed coatings, surface texturing, etc. on engine components is a way to reduce corrosion, friction, wear and weight of engines. In the paper, the potential of the application of HVOF sprayed coatings on the engine components is discussed with respect to their sliding friction properties. The technology of thermal spraying enables one to create the surface coating approximately 50 μm thick, which provides functional surface protection of the coated parts. HVOF technology offers the possibility of creating the coatings of materials based on the principle of hardmetals with high wear resistance and favourable sliding properties. Such a combination predestinates HVOF sprayed coatings for sliding applications, such as pistons of combustion engines, pumps and other hydraulic devices. In this application area they are used on a regular basis. In practice, the producers and users of thermally sprayed coatings face the problem of the interaction of the coatings and their counterparts with the presence of other media, fuels, or, in the case of sliding wear more often, lubricants. The paper describes the methodology of measurement suitable for lubricated HVOF coatings using the pin-on-disc test according to ASTM G-99 and on the evaluation of the influence of lubricants on the friction properties, wear rate and mechanism of HVOF sprayed coatings. The pin-on-disc test according to ASTM G-99 was performed on the thermally sprayed Cr_3C_2-25%NiCr coating to determine and describe their sliding wear behaviour under different test conditions. The influence of various test parameters (load, wear track diameter, temperature, lubrication, counterpart

WIT Transactions on Engineering Sciences, Vol 66, © 2010 WIT Press
www.witpress.com, ISSN 1743-3533 (on-line)
doi:10.2495/TD100121

material) on the coefficient of friction, wear mechanism and wear rate was monitored.

Keywords: HVOF, coating, sliding friction, ASTM G-99, pin-on-disc.

1 Introduction

The pin-on-disc wear test is one of the most widespread tests for sliding wear behaviour and friction of material pairs. Its basic configuration can be seen in Fig. 1. The arrangement of the testing equipment can be both horizontal and vertical, while vertical configuration enables to remove the wear debris during the test. A radius tipped pin is pressed against a flat disc. The relative motion between them causes a circumferential wear path on the disc surface is generated. Either the pin or the disc can be moving.

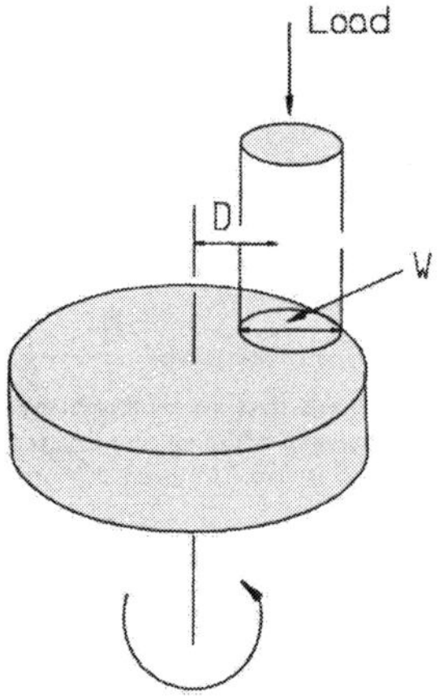

Figure 1: Diagram of the pin-on-disc test [1].

The test parameters that have been used vary. The ASTM standard for this test, ASTM G99, does not specify particular values for the parameters, but allows them to be selected by the user to provide the simulation of an application. The parameters can vary including size and shape of the pin, load, speed and material pairs. The test can be also performed in a controlled atmosphere and with lubrication [1]. The stress levels change during the test, as a result of the wear, and the relationship between wear and duration or amount of sliding is often non-linear. In order to make a material comparison, the ASTM standard recommends measuring the wear on both members after a fixed number of revolutions.

As a result of the pin-on-disc test, a mean value of coefficient of friction (CoF), CoF evolution in dependence on the number of revolution or sliding distance, and the wear of tested material can be considered. The wear is evaluated by measurement of material volume loss that could be determined by measurement of weight or by measurement of wear track profile.

2 Experimental

The thermally sprayed Cr_3C_2-25%NiCr material was used for sliding wear test measurements. The coating was sprayed by HP/HVOF JP-5000® (TAFA) spraying technology in the ŠKODA VÝZKUM s.r.o. in Plzeň, using the standard preparation procedure on the grit blasted substrate of carbon steel (ČSN 11 523) and the previously optimized spraying parameters [2].

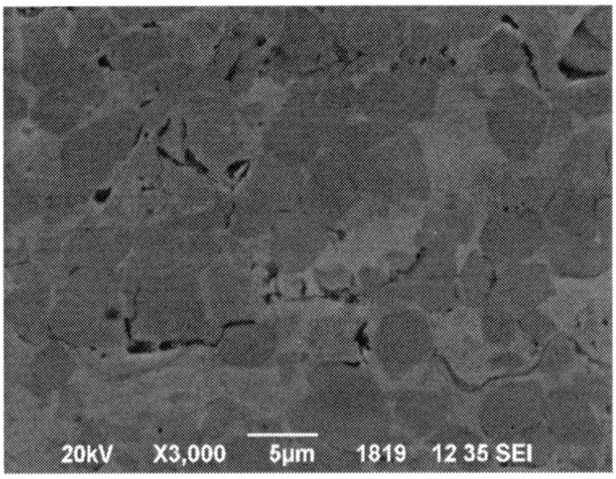

Figure 2: Cr_3C_2-NiCr coating microstructure.

The microstructure of the HVOF sprayed Cr_2C_3-25%NiCr coating (Fig. 2) is slightly heterogeneous, with less than 1% of porosity. A characteristic feature of the Cr_3C_2-NiCr coatings is the presence of numerous carbide grains bonded to the metallic matrix [3,4]. Besides the Cr_3C_2 carbides and the NiCr matrix the coating also contains Cr_2O_3 oxide, originated during spraying as a result of the oxidation process, and probably also the Cr_7C_3 and the $Cr_{23}C_6$ carbides of the hardness lower than the primary Cr_3C_2, originated in the process of de-carbonization during spraying similarly to the WC-Co coating [5]. The dissolution of carbides in the matrix leads to the increase in C and Cr concentration in the matrix compared to the powder composition. It enables the molten particles to solidify as the semi-amorphous or nanocrystalic C and Cr rich structure. The rapid solidification also contribute to the creation of other metal nanocrystalic phases with the grains size of cca 10 nm composed of 50% Ni and 50% Cr [6].

The basic mechanical properties of the tested Cr_3C_2-25%NiCr are summarized in Table 1. The value of indentation fracture toughness, stated in Table 1, was determined according to the Lawn-Fuller formula [7].

Table 1: Cr_3C_2-25%NiCr coating mechanical properties.

Coating	Thickness [µm]	Density [g.cm-3]	Microhardness HV0,3	Hardness HR15N	IFT (LF) [MPa.m-1/2]
Cr_3C_2-NiCr	433 ± 9	6.646	1030 ± 114	91 ± 0.8	1.45 ± 0.47

The coating sliding wear and friction properties were evaluated in the New Technology Center of University of West Bohemia in Plzeň (NTC ZČU), applying the pin-on-disc test according to ASTM G-99, using High Temperature Tribometer produced by CSM Instruments SA. The test was performed on the polished coating surface of the roughness Ra 0.16 ± 0.04. To study the influence of the test parameters on the sliding wear and friction coating properties the material of the pin, wear track diagonal, pin load and temperature varied from that usually used in NTC ZČU. The usually used parameters are: Pin: Al_2O_3 ball, ∅ 6 mm, Pin load: 10 N, Sliding speed: 0.1 m/s, Number of cycles: 50 000, Wear track diagonal: 2, 3.5 and 5 mm, Temperature: 20°C, Lubrication: dry. First, the measurement was performed by parental parameters. The influence of the wear track diagonal on the coating wear rate and the CoF dependence was evaluated. Then the material of the pin was altered by hardened steel, while for the other parameters the parental values were used. The influence of the pin load was evaluated by varying the pin load: 1, 2, 5 and 10 N. The influence of the wear test temperature was evaluated by means of varying the temperature of the test: 20°C, 500° and 700°C. The influence of lubrication on a mean value and evolution of CoF were evaluated using 3 types of engine oils: mineral MOGUL 15W-40, semi-synthetic MOGUL 10W-40 and synthetic MOGUL 5W-40.

Besides the studied parameter, the number of cycles altered due to the specific condition of each set of measurement.

3 Results and discussion

Because the CoF measurement using parental parameters served as a basis for all measurements sets, there is a lot of experimental data concerning the value and evolution of CoF of Cr_3C_2-NiCr coating, measured at the same test parameters during a long period of time. The overview of experimental results can be seen in Table 2.

Table 2: CoF values for the sliding pair Cr_3C_2-25%NiCr coating – Al_2O_3.

Experiment	2 mm wear track diagonal	3,5 mm wear track diagonal	5 mm wear track diagonal	Mean value
1	0.64 ± 0.06	0.66 ± 0.09	0.63 ± 0.08	0.64 ± 0.01
2	0.58 ± 0.06	0.59 ± 0.08	0.60 ± 0.07	0.59 ± 0.01
3	0.59 ± 0.06	-	0.57 ± 0.08	0.58 ± 0.01
4	0.66 ± 0.08	0.68 ± 0.09	0.68 ± 0.09	0.67 ± 0.01
Mean value	0.61 ± 0.03	0.64 ± 0.04	0.62 ± 0.04	0.62 ± 0.04

It can be stated, that the scatter of measured values is not high with the respect to fact, that the measured coatings sets were sprayed and measured during a long period of time (more than 5 years). The coating quality and also the accuracy of the tribometer were found reproducible. The evolution of the CoF curves in dependence on the number of cycles can be seen in Fig. 8. The CoF increased gradually to the stable value, which was reached at between 15 –

20 thousand cycles. The increase is connected with the high scatter of the CoF values. Similar CoF evolution was observed for all provided sets of experiments. The results can be found in the previous work [8–12] in more detail. The wear track diagonal did not influence the measured CoF in any set of measurement, so CoF could be considered independent of the wear track diagonal. Unfortunately, the same cannot be said about the wear rate. The wear loss of the hard wear-resistant Cr_3C_2-NiCr coating is too small to be measured. The measurable wear would be achieved by the increase in either pin load or a number of cycles, which is beyond the measurement range of the used tribometer. For reproducible evaluation of sliding wear it is necessary to choose some other type of the tribometer or a wear test, such as block-on-ring or crossed-cylinder wear test [1]. The major wear mechanism of the HVOF hardmetal coating is connected with the gradual primary loss of the metal matrix from the areas between coatings' hard particles, followed by the weakening of their attachment and pulling them out of the coating surface. In the case of the Cr_3C_2-NiCr coating, due to its lower fracture toughness, a higher amount of carbides and also bigger wear debris are pulled-out and cause the fluctuation of CoF curve. The design of the CSM Tribometer does not allow the wear debris to naturally fall off; they are trapped in the wear track and serve as an abrasive medium. Then the mechanism of wear changes from sliding to abrasive wear [2, 13]. In Figure 3b the area of a considerable damage of the wear track bottom, the weakened carbides and the deformed matrix can be seen. In [14] the cermet coatings the wear mechanism is described as the contact between the pin and coatings' carbides, which are slightly protruding from the matrix due to grinding. Such a condition leads only to a very slight wear. In [14] the formation of the tribofilm consisting of the oxidized, plastically deformed matrix was recognized, and could also be expected.

Figure 3: OM and SEM of wear track in the Cr_3C_2-NiCr coating.

The use of a hardened steel pin led to a significant increase in the middle CoF value (see Table 3) and to a change in the wear mechanism. The whole parts of the splats were pulled out in the case of the steel pin. Both effects could be caused by creation of micro-weld between pin material and NiCr matrix, or as it

Table 3: CoF values for the sliding pair Cr_3C_2-25%NiCr coating – hardened steel.

Wear track diagonal [mm]	2	3,5	5	Mean value
CoF	0.75 ± 0.06	0.61 ± 0.09	0.78 ± 0.08	0.71 ± 0.09

is mentioned in [15], the wear debris of the steel pin can serve as an abrasive medium that promotes the wear of Co binder in the coating and simultaneously cuts off the beneficial oxidic tribofilm.

The influence of the pin load on the CoF measured value of the sliding pair Cr_3C_2-NiCr - Al_2O_3 can be seen from the data in Table 4 and from the graph in Fig 5. The stable value of CoF is decreasing, together with the used pin load. The running-in period is shorter for a lower pin load. Both effects can be caused by the increasing impact of the surface roughness on CoF with the decreasing load. The influence is described in [11] in more detail.

Table 4: CoF values for the sliding pair Cr_3C_2-25%NiCr coating – Al_2O_3 – the influence of the pin load.

Pin load L [N]	10	5	2	1
CoF value	0.53 ± 0.01	0.55 ± 0.05	0.64 ± 0.04	0.67 ± 0.08

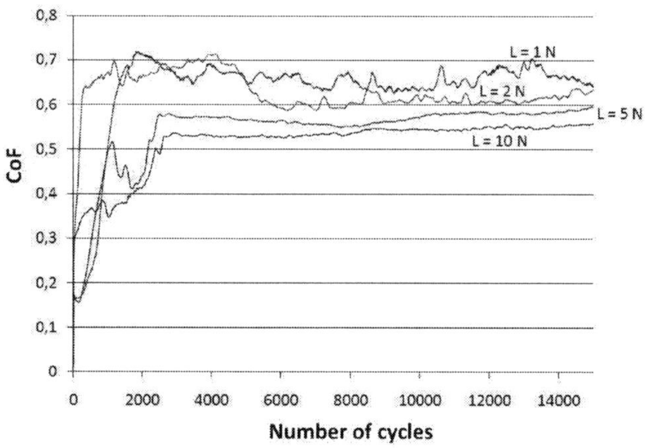

Figure. 5: The CoF evolution in dependence on the number of cycles – the influence of the pin load (smoothed curve).

The influence of a high temperature on the CoF values of the sliding pair Cr_3C_2-NiCr - Al_2O_3 can be seen from the data in Table 5 and from the graph in Fig 6. The highest CoF was measured at room temperature. At 500°C, the CoF was lower, its progression was steady, especially for 5 mm wear track diagonal. The lowest and most steady CoF was measured at the highest temperature, 700°C.

Table 5: CoF values for a sliding pair Cr_3C_2-25%NiCr coating – Al_2O_3 – the influence of a high temperature.

Temperature [°C]	20	500	700
CoF value	0.58 ± 0.07	0.47 ± 0.01	0.42 ± 0.01

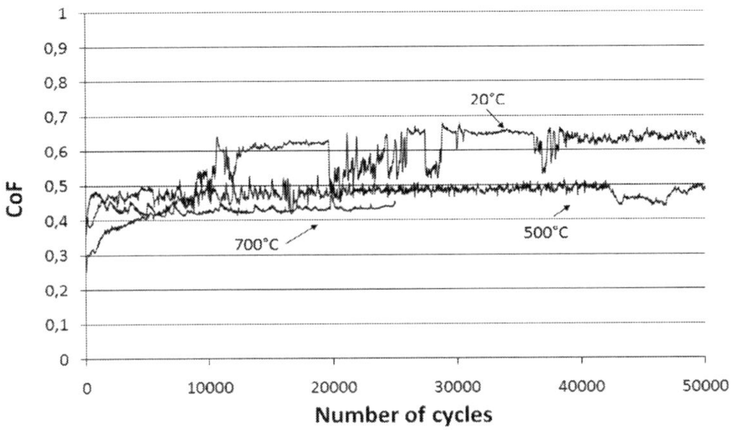

Figure. 6: The CoF evolution in dependence on the number of cycles – the influence of a temperature.

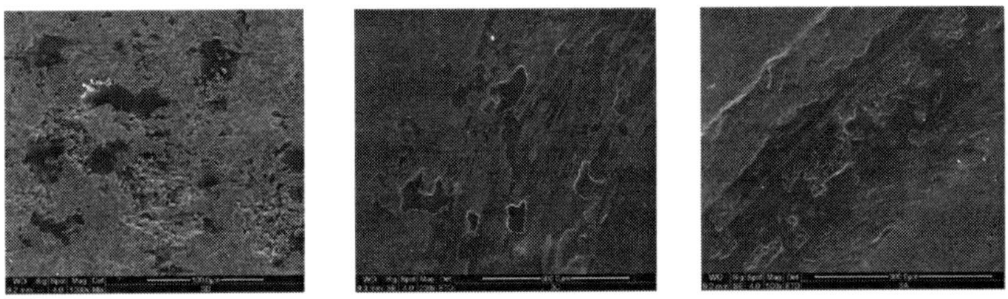

Figure 7: SEM of the Cr_3C_2-NiCr wear track at a) 20 °C, b) 500°C and c) 700°C.

At elevated temperature, an oxide layer consisting of Cr_2O_3 and $NiCr_2O_4$ [16, 17] appears. The Cr_2O_3 layer adheres strongly to the coating's surface and is characterized by CoF values between 0.25-0.5 [13], which is in a correlation with the CoF values measured in this study. According to [15, 16], the thickness and density of the oxide layer is different for 500°C and 700°C. For 700°C, it is thinner but denser, with better protective effect on the underlying surface. In [14] the existence of tribofilm on the Cr_3C_2-NiCr surface is described. In this study, its occurrence was proved by the EDX analyses of the oxygen amount in the wear track and on a free surface. The amount of oxygen absorbed in the wear track was higher compared to free surface for Cr_3C_2-NiCr coating, while for the other coatings it was the same. The influence of temperature on the friction properties of thermally sprayed hardmetal coatings is discussed in [2, 10] in more detail.

The influence of lubricants on the CoF values and evolution of the sliding pair Cr_3C_2-NiCr - Al_2O_3 can be seen from the data in Table 6 and from the graph in Fig 8. The character of the lubricated friction is completely different. Besides the significant decrease in the CoF mean value the scatter also decreases, the curves are smooth and the measurements in the wear tracks with the different wear track radius are almost similar. Part of the CoF evolution curve characterized by the CoF increase due to the creation of the wear track is not present; on the contrary, the stable value is lower than the initial one and is reached during the first 1000 cycles. The part of the CoF curve is probably connected with the creation of a stable film of the lubricant [13].

Table 6: CoF values for sliding pair Cr_3C_2-25%NiCr coating – Al_2O_3 – the influence of lubrication.

Wear track diameter	2 mm	3 mm	5 mm
Dry friction	0,658 ± 0,085	0,680 ± 0,093	0,677 ± 0,087
MOGUL 5W-40	0,119 ± 0,003	0,123 ± 0,002	0,123 ± 0,003
MOGUL 10W-40	0,120 ± 0,003	0,126 ± 0,003	0,126 ± 0,003
MOGUL 15W-40	0,115 ± 0,006	0,121 ± 0,003	0,121 ± 0,003

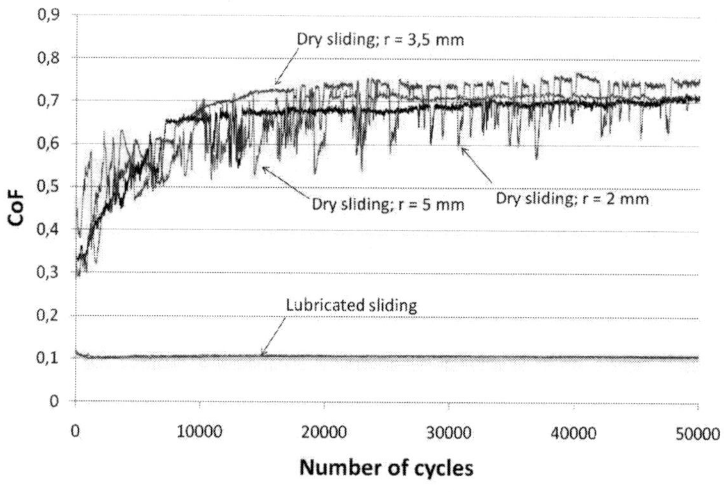

Figure 8: The CoF evolution in dependence on the number of cycles – the influence of lubrication.

The differences in values and character of CoF originated as a consequence of using different lubricants are very small, in terms of ASTM G-99 almost imperceptible. The lowest CoF was measured for MOGUL 15W-40, followed by MOGUL 5W-40 and MOGUL 10W-40.

The wear track bottom (Fig. 9) of the coating lubricated with MOGUL 5W-40 shows significantly lower wear damage, without the signs of the carbides loss. Only random cracks in the carbides and in the boundaries between the carbides and the matrix appeared. Comparing the SEM micrographs it can be said that the lubrication decreases the wear of the coated surface and it can be presumed that the wear rate, expressed by means of material volume loss, will be lower in the case of lubricated sliding, too. More information about the influence of lubricants can be found in [7].

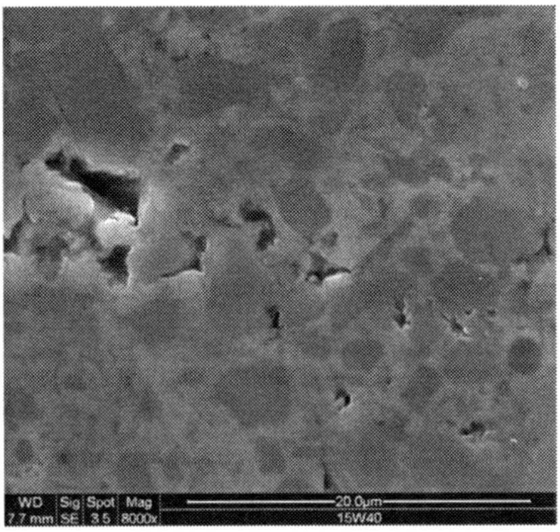

Figure 9: SEM of the Cr_3C_2-NiCr wear track performed under lubricated conditions.

4 Conclusion

To fully understand the friction and wear behavior of the HVOF sprayed Cr_3C_2-25%NiCr coating, the study of the influence of various pin-on- disc test parameters was performed with respect to the CoF mean values, the CoF dependence on the number of cycles and the wear mechanism. The wear rate of the coating was not evaluated due to a very small volume loss of hard, wear resistant coating, which is not reproducibly measurable. It was found that the hardened steel pin in the sliding pair with the Cr_3C_2-25%NiCr coating causes higher CoF compared to the sliding pair Cr_3C_2-25%NiCr coatig-Al_2O_3. The higher the used pin load, the lower the CoF in the case of Al_2O_3 pin. The high temperature leads to the creation of an oxide tribofilm in the wear track, which has better sliding properties than the original Cr_3C_2-25% NiCr surface. The CoF decreases with the increasing test temperature. The presence of the lubricant decreases the CoF value more than 5 times, the wear mechanism is changing to be less pronounced. The wear track under the lubricated condition is almost undistinguishable from the untouched surface.

Another important test parameter affecting the friction and wear properties is a sliding speed. Its influence will be studied in the near future.

Acknowledgement

The paper was written thanks to the financial support of the EUREKA E!4504 Project No. OE09001

References

[1] Bayer, R. G., Mechanical Wear Prediction and Prevention, Pub. Marcel Dekker, Inc, NY, USA (1994)

[2] Houdková, Š., Kašparová, M., Zahálka, F., Tribological behaviour of thermally sprayed coatings under high temperature, Proc. of Vrstvy a povlaky 2007, 29.-30.10.2007, Rožnov pod Radhoštěm, ISBN 978-80-969310-4-0, p. 49 (in Czech)

[3] Berger, L.M., Structure, Properties and Potentials of WC-Co, Cr_3C_2-NiCr and TiC-Ni based Hardmetal Like Coatings, In Proceeding of the 9th National Thermal Spray Conference 1996, ASM International, Materials Park, OH, USA, 1996, ISBN: 0-87170-583-4

[4] Berger, L.-M., Hardmetals as Thermal Spray Coatings, Powder Metallurgy, 50 (2007) 3, pp. 205-214.

[5] Berger, L.-M., Chemistry of carbide-metal composites for preparation of HVOF sprayed coatings, Papers, Materials Week 2001, Int. Cong. On Adv. Mater., their Processes and Applications, 1.4. October 2001, München, [CD-ROM]. 2002. p. 10. (ISBN 3-88355-302-6)

[6] Zimmermann, S., Kreye, H., Chromium carbide Coatings produced with Various HVOF Spray Systems, In Proceeding of the 9th National Thermal Spray Conf., Cincinnati, ASM International, Materials Park, Ohio, USA (1996), pp.147-152

[7] Lawn, B.R., Fuller, E.R., J. Material Science, 10, (1975), pp. 2016-2024

[8] Houdková, Š., Zábranská, I., Zahálka, F., The influence of lubricants on friction properties of HVOF sprayed coatings, suitable for combustion engines, In. Proc. Of 18th International conference on Metallurgy and Materials METAL 2009, May 19-21, Hradec nad Moravicví, Czech Rep., (2009), ISBN 978-80-87294-10-9, pp.258-265

[9] Houdková, Š., Zahálka, F., Kašparová, M., The influence of thermally sprayed coatings microstructure on their mechanical and tribological characteristics, Material Science Forum 2008, Vols. 567-568 (2007), pp 229-232.

[10] Houdková, Š., Zahálka, F., Kašparová, M., Berger, L.-M., Tribological behaviour of thermally sprayed coatings at elevated temperatures, In. Proc. of International Thermal Spray Conference (2008) June 2-4, Maastricht, The Netherlands, 2008, ISBN 978-3-87155-979-2, pp.1497-1502

[11] Houdková, Š., Bláhová, O., Tichotová, P., Novotná, K., Tribological characteristics of thermally sprayed coatings, In: Tribology of surface layers and coatings, proceeding of the 4th international tribology conference, (2004), June 17-18, Prague, Czech Republic, ISBN 80-239-3123-7, pp. 1-8

[12] Houdková, Š., Enžl, R., Bláhová, O., Abrasion wear rate evaluation of thermally sprayed coatings, Acta Metallurgica Slovaca, Vol. 10, no. 1(2004), pp. 792-794

[13] Bhushan, B., Introduction to Tribology, Pu. John Wiley & Sons, NY, USA, (2002)

[14] Bolelli, G., Cannillo, V., Lusvarghi, L., Manfredini, T., Wear behaviour of thermally sprayed ceramic oxide coatings, Wear 261,(2006), pp.1298-1315

[15] Guilemany, J.M., Miguel, J.M., Vizcaino, S., Climent, F., Role of tree-body abrasion wear in the sliding wear behaviour of WC-Co coatings obtained by thermal spraying, Surface and coatings Technology 140 (2001) 141-146

[16] Berger, L.-M., Woydt, M., Zimmermann, S., Schwier, G., Enzl, R., Tribological behavior of HVOF-sprayed Cr_2C_3-NiCr and TiC-based coatings under high-temperature dry sliding, Conf. Proc. Int. Thermal Spray Conf. & Exhibition ITSC 2004, 10-12 May, Osaka, Japan, (2004),CD (ISBN 3-87155-792-7), p.10

[17] Berger, L.-M. Zieris, R., Saaro, S., Oxidation of HVOF-sprayed hardmetal coatings, Conf. Proc. Int. Thermal Spray Conference & Exhibition ITSC 2005, 2-4 May 2005, Basel, Switzerland, CD (ISBN 3-87155-793-5), pp. 969-976

Investigation on contact surface damage of automotive connector by fretting corrosion

M. A. Beloufa
PRISME Institute EA 4229, ENSI de Bourges, France

Abstract

The main cause of electrical contact resistance degradation by corrosion is the vibration of contact interfaces. The purpose of this paper is to analyse the change of contact resistance under vibration test for uncoated sphere/plane contact made of new high copper alloys. The influence of electrical and mechanical properties of materials and mainly the hardness on contact resistance was studied in this work. During the fretting test, contact point was submitted to 16000 vibration cycles under fretting amplitude of 50 μm and 1 Hz frequency. The sphere part was fixed, while the plane part was submitted to a relative motion. At the end of the test, the fretted surfaces and the wear debris were analysed by scanning electron microscope and energy dispersive X-ray spectroscopy to evaluate damage, oxidation and elemental composition present in the wear surfaces. In addition, the measurement of the wear track profile using a 3D surface scanning system was introduced. Increases contact resistance and contact temperature has been examined during the fretting test. The results showed that the contact resistance for the harder alloy was higher than those obtained for the other materials. In addition, topographic measurement showed that the small wear track corresponds to the harder material. Finally, depending on the hardness and the resistivity, these uncoated materials influence fretting apparition and its level.
Keywords: contact resistance, EDX and SEM analysis, fretting corrosion, high copper alloys, wear surface, 3D scanning profiler, crack, delamination.

1 Introduction

Electrical connectors used in telecommunication, automotive and many other applications are exposed to degradation due to the vibration of mated contact surfaces. External mechanical vibration is the main cause of the fretting motion

WIT Transactions on Engineering Sciences, Vol 66, © 2010 WIT Press
www.witpress.com, ISSN 1743-3533 (on-line)
doi:10.2495/TD100131

between contacting surfaces which is manifested by the mechanical and electrical damage of contact interfaces (Fagerstrom and Nicotera [1]). The failure process begins by wear of contact surfaces then oxidation of these surfaces or of wear debris which can cause rapid and dramatic increases in electrical contact resistance, proceeding to virtual open circuits [2–4]. Low-level relative movement between mated contacts can also be induced by differential thermal expansion of materials (Bock and Whitley [5]).

Determination of contact resistance changes during fretting of coated contact materials has been the subject of many investigations [6–10]. Unfortunately, contact materials with different coating (silver, palladium, tin…) have many inconvenient. Silver is prone to tarnish in atmospheres containing certain sulphur or chlorine compounds, which limits its application in electronic connectors while palladium will catalyse the formation of insulating frictional polymers which originate in absorbed organic air pollutants on the fretting surface. Wear was responsible for the insulating materials at the interface which caused contact resistance to rise. When the contact finish is gold, a reduction in its wear is beneficial but gold is more expensive. The improvement due to the harder nickel under-plate has also been found. It has been shown that the thicker layer leads to decrease fretting contact resistance (Antler [6]). Commonly, the main degradation of electrical connectors due to oxidation is prevented by coating. Unfortunately, after a certain number of vibration cycle, the fretting wear removes this protective metal layer leading to exposure of the base metal substrate to fretting corrosion. Regarding the disappearance of protective coating, uncoated high copper alloys were studied in this work. The subject of this paper is to analyse the effect of these new materials in apparition and increases in fretting contact resistance. Fretting corrosion behaviour has been discussed and investigated in much paper using scanning electron microscope (SEM) and energy dispersive X-ray spectroscopy (EDX) [11–13]. In this study, oxidation of the contact surfaces and the wear debris was extended and confirmed with the EDX technique. Moreover, fretting damage on the contact zone by several wear mechanisms such as abrasion, delamination wear and accumulation of wear debris were observed with the SEM technique. Analysis of the contact surface profile with the help of a profiler has been the subject of numerous recent research studies [14, 15]. In this work, surface profile of the fretted zone was examined by a profiler and then the wear volume loss and the fretted contact area have been calculated.

2 Contact geometry

The samples with a U shape fig. 1 were made with a sheet of 20 mm width, using the processing technique of stamping and bending. The two parts of the sample (plane part or lower part and a sphere part or upper part) have the same thickness i.e. 0.8 mm. The radius of the sphere segment was 3 mm fig. 1. It has been demonstrated by [16, 17] that the contact resistance for sphere/plane contact was lower than the cylindrical/plane contact. Therefore, the tested U-shaped samples have a contact shape with spherical segment.

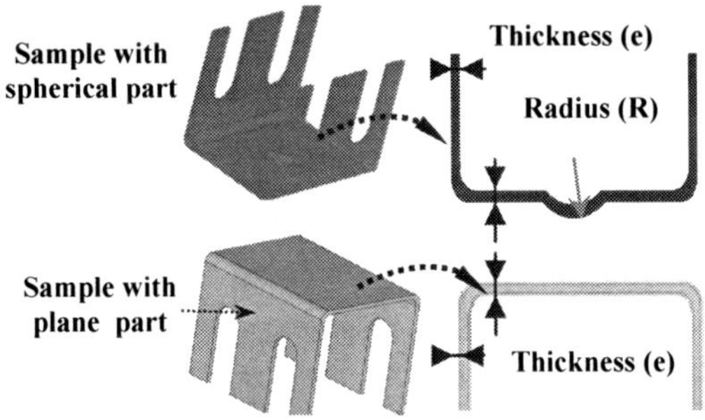

Figure 1: U-shaped samples.

3 Materials studies

The previous U-shaped samples were made by recent high copper alloys. Four types of contact materials have been selected for this study. These materials will be used in automotive power connectors and they combine excellent contact performance with good mechanical and electrical properties. Table 1 shows according to the descending order of electrical conductivity some of the electrical, thermal and mechanical properties of the selected alloys (Zauter and Kudashov[18]).

Table 1: Characteristics for different used copper alloys.

Copper alloy (UNS Designations)	C10100	C19210	C12200	C70250
Composition	Cu 99.99	CuFeP	Cu 99.9% + 0.008% P	CuNi3SiMg
Yield stress (MPa)	200	322	200	514
Vickers hardness (HV)	65-95	100-130	65-95	180-220
Young modulus (MPa)	$1.2\ 10^5$	$0.9\ 10^5$	$1.2\ 10^5$	$1.22\ 10^5$
Tensile strength (MPa)	240-300	360-440	240-300	620-760
Thermal conductivity (W/mK)	401	350	340	190
Electrical resistivity (Ω.mm)	$1.68\ 10^{-5}$	$1.88\ 10^{-5}$	$2.12\ 10^{-5}$	$4\ 10^{-5}$
Electrical conductivity (%IACS)	100	91	80	43

The primary selection criterion for these alloys was based on their variation on the electrical and mechanical properties. Table 1 shows that C70250 alloy is harder and less electrically conductive than the other copper alloys. So, the aim is to determine the relative susceptibility of these materials to fretting corrosion.

4 Test apparatus

Fig. 2 illustrates schematically a fretting test apparatus for measurement of electrical contact resistance during vibration.

The two parts of the contact specimen were fixed in the Teflon insulator supports. A DC power supply was used to apply a constant current which was equal to 2 A with an open circuit voltage of 2 V. This electrical current was transmitted from the flat sample to the spherical sample through the contact zone. When high currents are employed, there can be a profound effect on interfacial topography, film formation, contact resistance, resistance heating which accelerates oxidation, and micro-arcing.

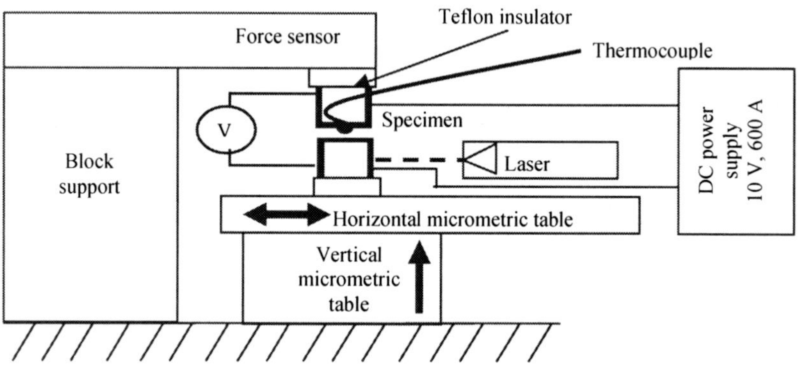

Figure 2: Schematic of the fretting apparatus and the circuit used to measure the contact.

A micro-voltmeter with 1 μV resolution was used to measure the voltage drop when passing a DC current through the contacts and then deduct the dynamic contact resistance. Contact resistance determinations were made using a DC four-wire dry circuit technique [19]. Constant contact force *Fc* was applied with the help of the vertical displacement of the micrometric table. This force was equal to 10 N and was controlled and maintained constant throughout the testing. The contact force was measured by a force sensor which was connected to a voltmeter through a conditioner. A horizontal micrometric table was used for producing fretting motion. These micrometric tables were driven by a DC stepping motor through a micrometer screw and were controlled by a motion controller. The upper sample or the spherical sample was fixed while the flat sample was submitted to a relative movement with a frequency of 1 Hz and fretting amplitude of 50 μm, this amplitude was measured by a laser with 0.1 μm resolution. The minimal distance between the tested flat sample and laser head was 3 cm. Bibliographical sources quote different values of displacement amplitudes as a border between fretting and reciprocating sliding motion. This value is variously interpreted and contained within a wide range of amplitudes between 50 and 300 μm (Ohmae and Tzukizoe [20]). Three specimens were used for each test. For each series of tests the fretting motion was carried for 16000 cycles where the contact voltage attained the open circuit value. The laboratory air ranged from 20 to 25°C and 40 to 50 percent relative humidity. Contact resistance and contact temperature data were collected by personal computer through GPIB General Purpose Interface Bus IEEE-488 and then stored for subsequent statistical analysis. Before each experiment, the samples surfaces were wiped with an antioxidant paste and then the samples were cleaned in an ultrasonic alcohol bath, dried by a wiper paper and carefully mounted in the fretting test assembly. Since the geometries of the flat and upper specimen were relatively small, thermo-couples with high response time (J type) was used to measure during the test the temperature variation near to the contact zone. The collected data of temperature were carried out with the help of a voltmeter through a conditioner. For security reasons, the test will be stopped when the contact temperature exceeds 50°C. In order to avoid any accidental arcing between the contacts, the connecting members were brought into contact with the current power supply OFF. The power supply was switched to ON only once the contact force was established and then the contact resistance and contact temperature were measured at each end of cycle and plotted according to the number of fretting cycles.

5 Experimental results and discussion

5.1 Electrical contact resistance

Fig. 3 shows the results with logarithmic scale of the contact resistance as function of fretting cycles only for high copper alloy C10100. The dotted curve presents the obtained results of contact resistance for three tested specimens. For all materials, the maximum contact resistance which corresponds to the open

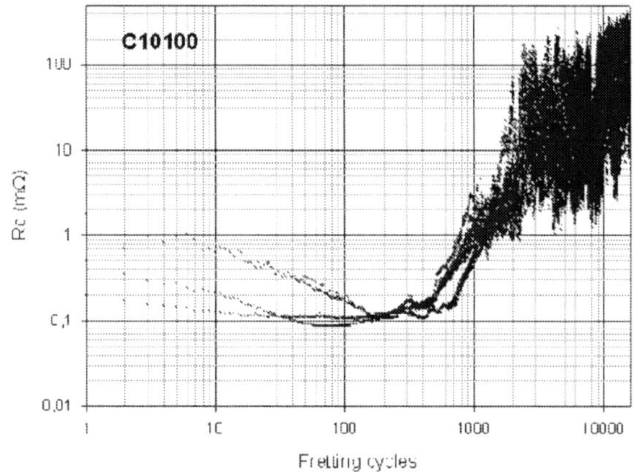

Figure 3: Change in contact resistance of copper alloy C10100 according to the fretting cycles.

circuit voltage of 2 V occurs at 16000 cycles where insulating materials tend to accumulate.

This is manifested by a rapid increasing of the contact resistance stability as the fretting progresses. For all materials, the contact resistance behaviour seems to be characterized by three distinct stages. In the first stage, which represents 500 cycles of fretting, the contact resistance decreases with very low fluctuation. Relative movement leads to a sliding/wiping action which facilitates the escape of the wear debris and cause burnishing with rupturing the asperities, ensures good metallic contact and increasing the conductive contact surfaces. Hence, the contact resistance decreases. It is well known that current passes through a micro asperities or spots present on the contact area. For metallic contact of rough surfaces [21, 22], has shown that contact resistance R_c can be approximated by the formula:

$$R_c = \frac{\rho}{2}\left(\frac{1}{na} + \frac{1}{\alpha}\right) \quad (1)$$

where ρ is the resistivity of contacting materials and α is the radius of a circle containing a cluster of n uniformly distributed contact spots, the radius of each spot is a. In the second stage, apparently between 500 cycles to 2500 cycles of fretting, the fretting was initiated and the contact resistance of all materials increases progressively until 2 orders of magnitude. However, as the fretting progresses, wear debris were generated and accumulated between the contacting surfaces and then oxidized. Removal of oxide exposes original surface which then oxidizes in a repeating process. Hard oxidized wear debris accelerates the progress of wear. As a result, contact resistance enhances continuously, this phenomenon was known as fretting corrosion. In the third stage or the saturation stage between 2500 cycles and 16000 cycles, the contact resistance increases sharply and is very unstable as characterized by large fluctuations between 1 and 3 orders of magnitude and sometimes by an open circuit. The peak and bottom values of contact resistance were speculated to be formed by accumulation of wear debris and oxidized wear debris in the fretting area, and subsequently, by extrusion of the accumulated wear debris on the contact area, respectively. Then, a hypothesis was set up that wear debris and copper-alloy have a different hardness from each other and that hard materials crush soft materials and crack the oxides which leads to a lower contact resistance. In this third stage, a thick insulating layer containing oxides and wear debris will form and any remaining metallic contact will be lost, causing a sharp increase in the contact resistance. In addition, prolonged exposure to fretting action will create the conditions for fatigue to occur and cause softening and progressive separation of contacting metals. When the current is confined to flow through the conducting spots (a-spots), the temperature of the point of contact (T_c) may be higher than that of the bulk (T_b). Hence, the increase in constriction résistance over the resistance that would exist (R^i_c) if the metal-metal contact were continuous across the entire contact area can be expressed as:

$$R_c = R^i_c\,[1 + 2/3(T_c - T_b)] \quad (2)$$

www.witpress.com, ISSN 1743-3533 (on-line)

The term ($T_c - T_b$) is called the super-temperature and is related to the voltage drop across the contact interface (U) as:

$$T_c^2 - T_b^2 = \frac{U^2}{4L} \tag{3}$$

where L is the Wiedemann-Franz Lorenz number (2.45 x 10^{-8} V^2/K^2). Hence, a small increase in the contact voltage (U) can raise the super-temperature high enough to produce metallurgical changes such as softening or even melting of the conducting areas (Timsit [23]). Accumulation of fretting debris and oxides in the contact zone reduces the size of contact spots surrounded by oxidized debris thus increasing the current density and the super-temperature of these conducting paths (Gagnon and Braunovic [11]). The contact temperature measured after 16000 cycles for all materials was about 30°C.

5.2 Comparison of contact resistance results

Fig. 4 shows a comparison between the smoothing curves of contact resistance for the all tested materials. A running average smoother was applied with sampling proportion of 0.01 and an interval of 100 points. Contact resistance was slightly higher for the harder material C70250 then the other softer

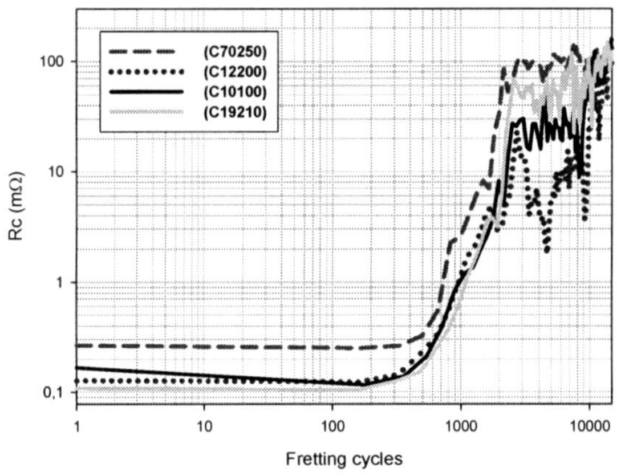

Figure 4: Comparison of smoothing contact resistance results for tested materials.

materials. It has been shown in an earlier paper (El Abdi et al. [24]) that static contact resistance measured under indentation test for C70250 alloy was higher then the other alloys. C70250 copper alloy which contains hard Nickel element has been hardened by precipitation and offer an interesting mechanical properties with high strength, high Young's modulus, high hardness and a low electrical conductivity (table 1) (Zauter and Kudashov [18]). This causes the highest contact resistance to be obtained with lower contact area.

6 Surface studies of fretted contact zone

6.1 SEM observation and EDX analysis

Fretting damage of contact surfaces was examined by Scanning Electron Microscopy (SEM). Some representative examples of the SEM examination of the fretting damage in contact zones are shown in fig. 5 and fig. 6 SEM micrographs were taken at 16000 cycles before and after cleaning of the fretted surface for the harder material C70250 and softer material C12200. Fig. 5 (a) shows the accumulated wear debris at the edges of the fretted zone for the softer alloy C12200 and the zoom of wear particles with x50000 of magnification. Wear debris seem to be more compact than the observed wear debris for the harder alloy C70250 (fig. 6(a)).

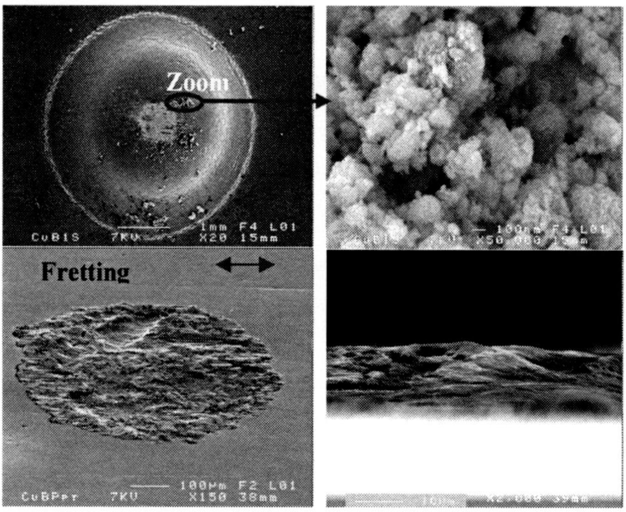

(a) Spherical sample before cleaning of the fretted surface.

(b) Flat sample after cleaning of the fretted surface.

Figure 5: SEM micrograph of the fretted contact zone for C12200 copper alloy.

Fig. 5(b) shows delamination wear and wears marks following the direction of the fretting motion caused by the abrasive particles remaining between the two contacting areas. A pushed material in the edge of the wear track was observed on the cross sectional fig. 5(b). The characteristic features of plastic deformation, abrasion and delamination wear were observed in the contact zones (fig. 6(b)). The micrograph of the cross sectional presented in fig. 6(b) shows a surface degradation which is characterized by extensive delamination and cracking as a result of contact fatigue (Antler [25]). It is clear that fretting causes serious damage in the contact areas. Typical damage being severe as abrasion, delamination and accumulation of the wear debris in the edges of the wear track leads to an increase of contact resistance. Elemental compositions of the fretted contact zones were determined by the Energy Dispersive X-rays analysis (EDX).

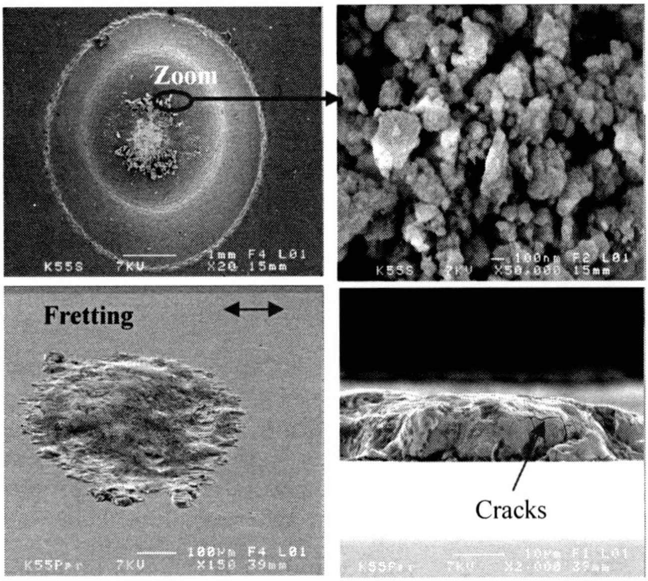

(a) Spherical sample before cleaning of the fretted surface.

(b) Flat sample after cleaning of the fretted surface.

Figure 6: SEM micrograph of the fretted contact zone for C70250 copper alloy.

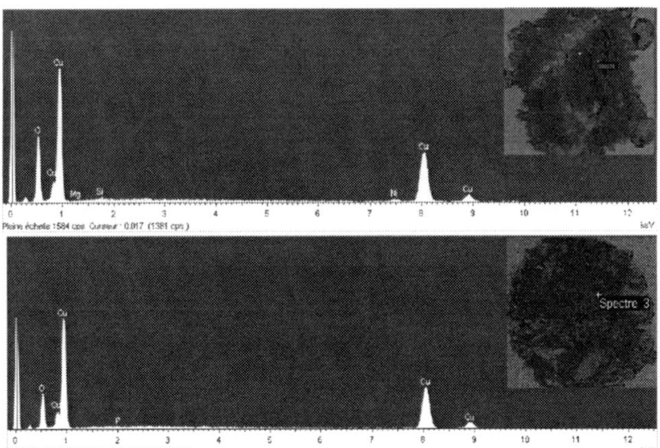

(a)

(b)

Figure 7: EDX spectra taken inside the fretted surfaces. (a) C70250 copper alloy; (b) C12200 copper alloy.

The main advantage of this type of X-ray fluorescence analysis is the ability to identify a wide range of elements simultaneously and to obtain a semi-quantitative measure of their concentrations in the contact regions under examination. Fretting contact areas have been examined by EDX analysis after the end of 16000 fretting cycles inside the wear track and on the wear debris for the C12200 and C70250 alloys. The elemental composition spectre measured from a spot across the wear track for the two materials was shown in fig. 7. The spectre presented in fig. 7 (a) shows the presence of strong peak of oxygen (26.5%) for the harder alloys C70250 and 20.7% of oxygen for the softer material C12200 (fig. 7 (b)). The presence of strong peaks of oxygen and copper indicates that wear tracks were composed mainly of these elements.

X-ray elemental mapping clearly reveal the nature of chemical species present at the contact zone and the extent of oxidation. The X-ray elemental mapping confirms the oxidation of the removal copper.

6.2 Profile measurement

The cleaned wear track of the tested alloys was examined by a 3D scanning profiler [26] to evaluate the dimensions of the wear tracks. Fig. 8 shows the smoothing initial profiles and the smoothing profiles of the wear tracks for the C70250 and C12200 alloys obtained after 16000 fretting cycles. The X,Y grid resolution for the collected data was 1.2 μm over a measurement area of 1.2 mm x 1.2 mm. 3D sub-micrometric profiler method can determine the parameters of wear in fretted surfaces.

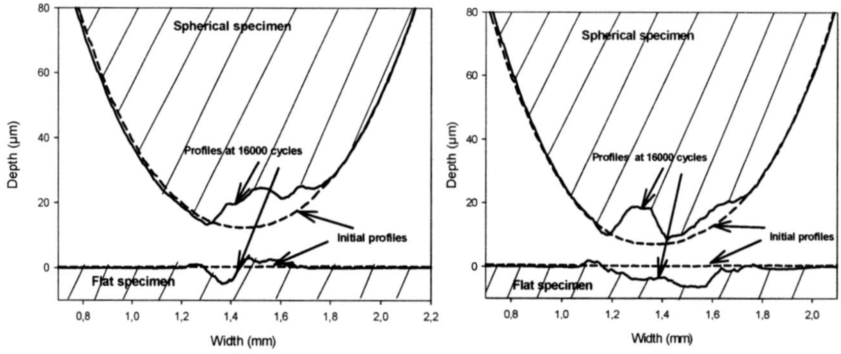

(a) (b)

Figure 8: Scanning fretted surfaces and profiles of the wear track after 16000 fretting cycles (a) C70250 copper alloy; (b) C12200 copper alloy.

The real contact area, volumetric wear removal, width, and length of the wear track were presented in table 2 for the flat specimen and for all tested materials.

Table 2: Wear sizes for all tested high copper alloys.

	C10100	C12200	C19210	C70250
Depth (μm)	7.7	6.54	7.8	4.94
Width (μm)	593	600	659	450
Fretted area (mm^2)	0.37	0.34	0.35	0.28
Fretted volume loss (mm^3)	5.9 10^{-4}	4.7 10^{-4}	6.1 10^{-4}	1.4 10^{-4}

The measured wear sizes for the harder material C70250 were lower than the softer materials C10100, C12200 and C19210. The softer materials suffer more surface damage, material loss, and generation of wear debris.

The harder materials suffer very little surface damage and not much wear debris generation. It is well known that the wear volume is proportional to load F_c and fretting amplitude x and is inversely proportional to hardness H, the wear volume V can be calculated by the following equation of Archard [27]:

$$V = \frac{2\,K\,F_c\,N\,x}{3H} \tag{4}$$

where N is the total number of cycles, and K is a wear coefficient.

The radius of contact area (a) can be calculated by the following equations:

$$a = \left(\frac{3F_c R}{4E^*}\right)^{1/3} \tag{5}$$

$$\frac{1}{E^*} = \frac{2(1-\nu^2)}{E} \tag{6}$$

Where R is the radius of spherical specimen (R=3mm), E is the Young modulus and ν is the Poisson coefficient. The calculated diameter by the eqns. (5) and (6) for the C12200 alloys was equal to 140 µm. Fig. 9 gives a schematic

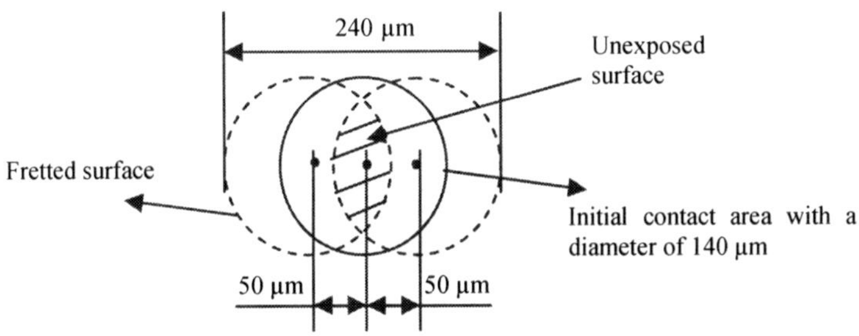

Figure 9: Fretted surface for one fretting cycle (fretting amplitude of 50 µm, C12200 alloys).

of the fretted surface for one fretting cycle. The fretted area after 16000 fretting cycles was higher than the fretted area for one cycle. The measured width of the track for the C12200 after 16000 fretting cycle was equal to 600 µm (table 2) and was higher than the initial width 240 µm. The width and depth increase during the fretting test. This is due to the progressive wear of the spherical specimen; the initial contact area was spherical and then increase and became flat.

7 Conclusion

Resistance increases noted during fretting action are felt to be due to trapped wear debris composed mainly of oxides. Based on the variation in contact resistance as a function of fretting cycles, wear debris, elemental composition of the wear track, surface profile and wear track sizes, were assessed by surface analytical techniques (SEM, EDX and profile-meter). SEM results showed that fretting causes serious damage in the contact areas as abrasion, delamination wear, fatigue cracks and accumulation of the oxidized wear debris which can leads to an increase of contact resistance. EDX analysis confirmed the presence of a strong percentage of oxygen in the wear track for the tested copper alloys. 3D scanning profiler showed that softer materials suffer more surface damage, material loss, and generation of wear debris than the harder material. The hard material was more wear resistant than soft materials and lead to high electrical contact resistance during fretting test. These studies underline the importance and necessity of using different surface characterization techniques in order to obtain an unambiguous explanation and information of the phenomena occurring at the contacting interfaces. The obtained results could help the connector manufacturers in the choice of contact materials and were a useful tool to evolve compromise between the electrical and mechanical aspects for connectors.

Some proposed solution could lead to a decreasing of the contact resistance and increasing the contact durability as the use of a lubricant between the two contacting surface and increasing of the contact load which will reduce the micro-displacement of the contact interface by increasing the friction force.

References

[1] Fagerstrom, W.B., Nicotera, E.T., Fretting corrosion in connectors. *Proc. Connectors and Interconnections Symposium. Electronic Connector Study Group*, Philadelphia, pp. 303–312, 1961.

[2] Sauger, E., Fouvry, S., Ponsonnet, L., Kapsa, Ph., Martin, J.M., Vincent, L., Tribologically transformed structure in fretting. *Wear*, **(245)**, pp. 39–52, 2000.

[3] Braunovic, M., Fretting damage in tin-plated aluminium and copper connectors. *IEEE Transactions on Components, Hybrids, and Manufacturing Technology*, **(12)**, pp. 215–223, 1989.

[4] Malucci, R.D., Fretting corrosion degradation, threshold behavior and contact instability. *Proc. of the 49th Int. Conf. On Electrical Contacts IEEE Holm*, Washington, pp. 1–15, 2003.

[5] Bock, E.M., Whitley, J.H., Fretting corrosion in electric contacts. *Proc. of the 20th Int. Conf. On Electrical Contacts IEEE Holm*, Chicago, pp. 128–138, 1974.

[6] Antler, M., Survey of contact fretting in electrical connectors. *Proc. of the 30th Int. Conf. On Electrical Contacts IEEE Holm*, Chicago, pp. 1–21, 1984.

[7] Noël, S., Lécaudé, N., Correia, S., Gendre, P., Grosjean, A., Electrical and tribological properties of tin plated copper alloy for electrical contacts in relation to intermetallic growth. *Proc. of the 52nd Int. Conf. On Electrical Contacts IEEE Holm*, Montreal, pp. 1–10, 2006.

[8] Ito, T., Takata, K., Hattori, Y., Iida, K., Saitoh, Y., Microscopy study of fretting corrosion of tin plated contacts. *Proc. of the 53rd Int. Conf. On Electrical Contacts IEEE Holm*, Pittsburgh, pp. 216–221, 2007.
[9] Park, Y.W., Sankara Narayanan, T.S.N., Lee, K.Y., Fretting corrosion of tin-plated contacts. *Tribology International*, **(41)**, pp. 616–628, 2008.
[10] Van Dijk, P., Rudolphi, A.K., Klaffke, D., Investigations on electrical contacts subjected to fretting motion. *Proc. 21st Int. Conf. On Electrical Contacts (ICEC)*, Zurich, pp. 1–7, 2002.
[11] Gagnon, D., Braunovic, M., Effect of fretting in lead-free systems. *Proc. of the 50th Int. Conf. On Electrical Contacts IEEE Holm*, Seattle, pp. 248–254, 2004.
[12] Braunovic, M., McIntyre, N.S., Chauvin, W.J., Aitchison, I., Surface analysis of fretting damage in electrical contacts of aluminum with different contact materials. *Proc. of the 29th Int. Conf. On Electrical Contacts IEEE Holm*, Chicago, pp. 231–242, 1983.
[13] Kassman, A., Jacobson, S., Surface damage, adhesion and contact resistance of silver plated copper contacts subjected to fretting motion. *Wear*, **(165)**, pp. 227–230, 1993.
[14] McBride, J.W., Developments in fretting studies applied to electrical contacts. *Proc. of the 52nd Int. Conf. On Electrical Contacts IEEE Holm*, Montreal, pp. 170–180, 2006.
[15] McBride, J.W., The loaded surface profile: A new technique for the investigation of contact surfaces. *Proc. 23rd Int. Conf. On Electrical Contacts (ICEC)*, Sandai, pp. 150–156, 2006.
[16] Beloufa, A., Influence of shapes, contact forces and high copper alloys on the contact resistance and temperature. *Proc. 2nd Int. Conf. On Engineering Mechanics, Structures, Engineering Geology (EMESEG '09-WSEAS)*, Rodos Island, pp. 139–144, 2009.
[17] Queffelec, J.L., Benjemaa, N., Travers, D., Pethieu, G., Materials and contact shape studies for automobile connectors development. *Journal of IEEE Transaction*, **(14)**, pp. 90–94, 1991.
[18] Zauter, R., Kudashov, D.V., Precipitation hardened high copper alloys for connector pins made of wire. *Proc. 23rd Int. Conf. On Electrical Contacts (ICEC)*, Sandai, pp. 257–261, 2006.
[19] Standard Methods for Measuring Contact Resistance of Electrical Connections (Static Contacts), B53g-80, American Society for Testing and Materials, Philadelphia.
[20] Ohmae, N., Tzukizoe, T., The effect of slip amplitude on fretting. *Wear*, **(27)**, pp. 281–293, 1974.
[21] Holm, R., *Electric Contacts: Theory and applications*, Springer-Verlag: Berlin, 1999.
[22] Greenwood, J.A., Constriction resistance and the real area of contact. *Journal of Applied Physics*, **(17)**, pp. 1621–1632, 1966.
[23] Timsit, R.S., On the evaluation of contact temperature from potential-drop measurements. *IEEE Transactions on Components, Hybrids, and Manufacturing Technology*, **(6)**, pp. 115–121, 1983.

[24] El Abdi, R., Beloufa, M.A., Benjemaa, N., Contact resistance study of high copper alloys used in power automotive connectors. *Journal of Automobile Engineering*, **(222)**, pp. 1375–1383, 2008.
[25] Antler, M. Sliding wear of metallic contacts. *Proc. of the 26th Int. Conf. On Electrical Contacts IEEE Holm*, Chicago, pp. 1–24, 1980.
[26] TaiCaan Technologies Ltd, http://www.taicaan.com.
[27] Archard, J.F., Contact and rubbing of flat surfaces. *Journal of Applied Physics*, **(24)**, pp. 970–981, 1953.

Dressing monitoring with a digital signal processor

R. F. Luzia[1], P. R. de Aguiar[1], E. C. Bianchi[2] & M. M. Spadotto[1]
[1]*Department of Electrical Engineering, UNESP - Univ Estadual Paulista, Brazil*
[2]*Department of Mechanical Engineering, UNESP - Univ Estadual Paulista, Brazil*

Abstract

Grinding is a finishing process in machining operations, and the topology of the grinding tool is responsible for producing the desired result on the surface of the machined material. The tool topology is modeled in the dressing process and precision is therefore extremely important. This study presents a solution in the monitoring of the dressing process, using a digital signal processor (DSP) operating in real time to detect the optimal dressing moment. To confirm the monitoring efficiency by DSP, the results were compared with those of a data acquisition system (DAQ) and offline processing. The method employed here consisted of analyzing the acoustic emission and electrical power signal by applying the DPO and DPKS parameters. The analysis of the results allowed us to conclude that the application of the DPO and DPKS parameters can be substituted by processing of the mean acoustic emission signal, thus reducing the computational effort.
Keywords: data acquisition, digital signal processor, dressing operation, grinding process.

1 Introduction

Grinding is a workpiece finishing process that accounts for approximately 25% of all machining processes. Because this operation is usually the last one in the manufacturing sequence, all the time and quality-related problems generated along the chain of processes involved in the fabrication of a component must be

WIT Transactions on Engineering Sciences, Vol 66, © 2010 WIT Press
www.witpress.com, ISSN 1743-3533 (on-line)
doi:10.2495/TD100141

solved during grinding. As a result, any benefit obtained in grinding is of high value to the entire process [1].

Continuous grinding causes the grinding wheel to undergo surface wear and assume an inadequate shape, causing the appearance of dull edges, i.e., loss of cutting edges, as well as clogging of its pores by chips, making the removal of new material inefficient [2].

This grinding wheel surface wear reduces the tool's cutting ability, which in turn increases the temperature of the grinding process due to interaction between the cutting energy and the workpiece surface. Increases in temperature may lead to thermal damage, which may vary from a superficial change in color to fracture of the workpiece [3].

Analyzing the operating conditions and the state of the cutting tool enables one to predict the tool's service life. However, monitoring of the grinding process is highly complex due to the large number of parameters involved [4].

Focusing on the need to monitor and improve the grinding process, Webster et al. [5] studied the emission of acoustic signals from the workpiece surface and found that the so-called raw signal (AEraw) reliably represents the existence of faults on the surface of the workpiece and grinding wheel. Based on these studies, it was also concluded that the effective acoustic emission signal (AErms) can be used in the process for the same purpose, but, according to the integration time constant, some important data may be lost due to the characteristic of low pass filtration in this type of operation.

Aguiar et al. [6] proposes a combination of the effective acoustic emission signal and the mean electrical power signal of the grinding wheel motor, in a statistical tool called DPO to detect the occurrence of burn during the grinding process. By means of a second statistic studied by Dotto et al. [7], called DPKS, it is possible to identify the beginning and end of burn in most cases. This statistic also uses the same signals as DPO, and was designed to present the highest possible sensitivity in the process.

Because the effect of burn on the workpiece results from the wear of the grinding wheel surface, the occurrence of this effect can be reduced in the preparation of the grinding wheel prior to the grinding. The dressing operation prepares the grinding wheel, restoring the cutting edges and removing the chips from the grinding wheel, thereby reducing the temperature at the point of contact between the workpiece and the grinding wheel and improving the grinding process. When correct dressing parameters and conditions are used, it is possible to increase the ool's service life [3].

Aiming to improve the dressing process, Hellmeister [8] proposes the use of the raw acoustic emission signal in this process and concludes that by analyzing this signal it is possible to detect faults on the grinding wheel surface during dressing, which is in agreement with what Webster et al. [5] found in the grinding process.

Aguiar et al. [9] proposed the use of DPO and DPKS statistics in the dressing process, both of which presented satisfactory results. An analysis of the results of the statistics allowed for the determination of the moment when dressing could be concluded.

This article proposes the implementation of data acquisition and DPO and DPKS statistics processing routines on DSP hardware to ascertain the possibility and feasibility of developing a device operating in real time to monitor the dressing process.

After carrying out tests and analyzing the data, it was found that the DSP system showed satisfactory results, and that it is feasible to develop a tool based on this type of processing to monitor the dressing process. Moreover it was found that for the conditions applied, the calculated statistics can be substituted for the mean effective acoustic emission signal, which reduces the computational effort and simplifies the hardware.

2 Monitoring of the dressing process

The topology of the grinding wheel is defined by the spatial distribution and shape of the abrasive grains on the surface of the tool. Before beginning the grinding process, the grinding wheel must be prepared, i.e., material must be removed from the cutting surface with the least possible macroscopic alteration, and the *dressing* stage consists of conditioning the grinding wheel with the purpose of imposing a given grinding behavior. In conventional abrasive grinding wheels, these stages are carried out by the same process, which is normally called dressing [10]. In general, the dressing operation reconditions the surface of the grinding wheel so as to recover its cutting edges, which have been deteriorated by use, removing chips and allowing for the removal of new material [11].

According to Marinescu et al. [12], the process of preparing the grinding wheel is very important because the tool's topology is directly associated with its grinding performance through the characteristics of the forces acting upon the system, i.e., power consumed, process temperature, and the final surface of the ground workpiece. The choice of the values of the parameters to be used in the dressing process depends on the desired characteristics and on the desired final result of the grinding process; hence, the dressing is defined by the effect of grinding.

In vitrified conventional abrasive grinding wheels, the dressing tool is called a dresser. The most common dressers are single or multiple-tip diamond dressers with transverse movement [12, 13]. Dresser diamonds may be synthetic or natural and are available in several shapes and configurations, with some of their specification parameters being the type and way they are mounted in the tool, and the crystal's number, size and geometric characteristics.

The action of the dresser is similar to that of a turning tool and the volume of abrasive material removed from the grinding wheel in each pass is proportional to the dressing depth (a_d) [12]. The advancement of the dressing tool across the rotating grinding wheel causes the formation of a screw whose step is called a dressing step (f_d), which is given by:

$$f_d = \pi \cdot d_s \cdot v_{fd} / v_s \tag{1}$$

where d_s is the initial diameter of the grinding wheel, v_{fd} is the transverse advancement speed of the dresser, and v_s is the tangential speed of rotation of the grinding wheel.

Linked to parameter a_d is the width of action of the dresser (b_d), which defined the width of contact between the dresser and the grinding wheel [12]. Although the efficiency and quality of the dressing process depend mainly on a_d and f_d, the most important parameter is the overlap factor (U_d), which expresses a relationship between b_d and f_d.

$$U_d = b_d / f_d \tag{2}$$

The adjustment of parameters a_d and f_d defines the type of dressing [12]. Fine dressing is when both these values are low, and in this type of dressing, the distance between grains is small and grain density on the surface is high, causing the volume of material removed in grinding to be greater, with high initial forces and power that decrease with the wear of the grinding wheel.

Analogously, coarse dressing is when a_d and f_d have high values. Contrary to fine dressing, coarse dressing increases the distance between the abrasive grains, reducing the density, which in turn reduces the forces and power, although the roughness of the ground surface is greater.

Köning apud Aguiar [9] report that due to the wear that occurs at the tip of the dresser, adjustment of the process based solely on a_d and f_d is inadequate, since it does not consider the width of action, which varies with use. Therefore, the U_d parameter is more suitable for the adjustment of the dressing process, because it takes into account the real width of action according to the wear at the tip, making the process more efficient.

According to Xue apud Aguiar [9], the number of dressing passes depends on the intensity of wear of the grinding wheel, the grain size, and the depth of wear to be removed. Normally two or three successive passes eliminate the effects of previous dressing and grinding. However, in the dressing process errors may occur that affect the quality of the preparation of the grinding wheel in different ways. These errors may result from variations in dressing depth, dresser wear and other problems. Thus, a monitoring system for the process may detect such errors, as well as determine the moment when dressing should be concluded.

Xue et al. [22] use the acoustic emission signal to monitor the dressing process online in order to detect grinding wheel deformations and contour errors. This signal has proved efficient in the identification of faults that may occur in the process. Also, according to Xue et al., monitoring the dressing parameters results in a more homogeneous surface roughness and online monitoring can reveal the dressing depth behavior.

Due to the importance of the grinding process, numerous studies have been carried out to improve it. For example, acoustic emission sensors have been utilized to detect the contact between the grinding wheel and the dresser (Oliveira et al. [14]; Karpuschewski et al. [4]), monitor the dressing depth (Inasaki [15]), monitor the dresser lead (Inasaki [15]; Webster et al. [16]), diagnose dressing failures (Karpuschewski et al. [4]; Konig and Meyen [17]; Kwak and Ha [18]; Lee et al. [19]), model the dressing process (Linke [20]), finish dressing (Aguiar et al. [9]), and a method without contact has been proposed (Young and Chen [21]).

Because it is not possible to determine the number of passes required to complete the dressing process, Aguiar et al. [5] use DPO and DPKS statistics to monitor the process and determine the optimal dressing moment, i.e., when the process can be concluded. These statistics are used in the detection of the occurrence of the burn phenomenon in the grinding process.

According to studies carried out by Aguiar et al. [6], the combination of the effective acoustic emission signals (AE_{rms}) and the mean electrical power provide an excellent indicator of the occurrence of burn. This combination gave rise to the DPO parameter, which is given by the multiplication of the standard deviation of the acoustic emission signal by the maximum value of the mean electrical power signal recorded in the pass.

$$DPO = \sigma_{EA_{rms}} \cdot MAX(\overline{P}) \tag{3}$$

where σ_{EArms} is the standard deviation of the effective acoustic emission signal in the pass and *MAX(P)* is the maximum mean electrical power in the pass.

Modifying the expression of the DPO parameter in order to achieve greater sensitivity, the DPKS parameter was developed, which is given by multiplying the standard deviation of AE_{rms} by the sum of the difference between the instantaneous value of the mean electrical power and the standard deviation of this power raised to the 4th power [7].

$$DPKS = \sigma_{EA_{rms}} \cdot \sum_{i=1}^{i=m} \left(\overline{P}_i - \sigma_{\overline{P}}\right)^4 \tag{4}$$

where σ_P is the standard deviation of the mean electrical power signal in the pass, and P_i is the instantaneous value of the mean electrical power.

3 Methodology

This study consisted of analyzing the possibility of implementing a monitoring system for the dressing process using a digital signal processor (DSP) model ADSP-BF533 of Analog Devices' BlackFin family operating in real time. The analysis was performed by comparing the processing results of the DSP system in real time against the offline processing results in routines executed with MATLAB software. Os signals used in the MATLAB processing were obtained by a data acquisition system (DAQ), using LabVIEW software and a National Instruments signal acquisition board.

The DSP system was programmed to capture the mean quadratic acoustic emission signal (AERMS) and the mean electrical power (P) of the grinding wheel motor by means of a digital-to-analog converter. The acoustic emission signal was obtained by a passive response sensor at a frequency of 50 Hz to 1 MHz and preprocessed by a DM-42 acoustic emission module. The acoustic emission module receives the acoustic emission signal from the sensor and processes the RMS value with a time constant of 1 ms. To prevent saturation and

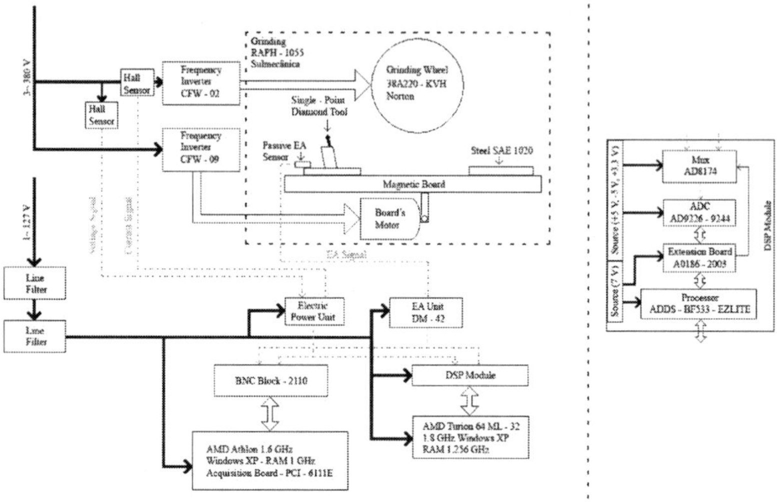

Figure 1: Schematic diagram of the test bench.

distortion in the conversion of the signal by the digital-to-analog converter used by the DSP, the acoustic emission module was adjusted to an input gain of 10 and an output gain of 20.

The electrical power signal was captured by a power module that uses two Hall effect sensors to take voltage and current readings. The module computes the voltage and current signals, supplying the instantaneous and mean power, and is therefore interesting in the study of only the mean power.

Figure 1 presents a schematic diagram of the test bench.

The experimental part consisted of six process tests, each of which comprised six dressing passes. A different overlap factor was used in each test. Because the dressing depth value was set to a single value, the variation of the overlap factor was determined by the variation of the transverse speed of the table.

By choosing the desired overlap factor and knowing the width of the dresser's action, the rotation of the grinding table motor can be determined by (5).

The width of the dresser's action was measured with a profile projector. This procedure was carried out before each test, since the dresser undergoes wear each time it is used, causing the value bd to vary even when ad is kept constant.

$$n = \left(\left(b_d \cdot 30 / U_d\right) + 0{,}48\right) \cdot 30 / 0{,}224 \quad (5)$$

where n is the rotation parameter given in rpm which must be adjusted in the frequency inverter that controls the table's motor.

In order to produce wear on the grinding wheel before each test, several grinding passes were performed at a cutting depth of 20 μm without using cutting fluid, until the burn effect occurred on a steel 1020 workpiece.

Table I lists the values of the parameters used in the tests.

Table 1: Test parameters.

Test	a_d (µm)	U_d	b_d (µm)	V_d (mm/s)	n (rpm)
1st	40,0	3,5	231,0	1,980	291
2nd	40,0	3,0	238,0	2,375	347
3rd	40,0	2,5	234,0	2,808	409
4th	40,0	2,0	233,0	3,498	507
5th	40,0	1,5	234,0	4,674	675
6th	40,0	1,0	235,0	7,050	1013

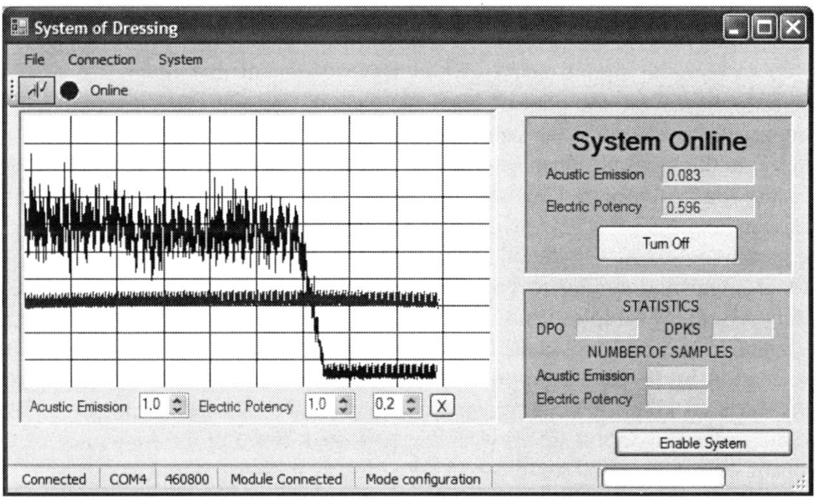

Figure 2: Online operation window with visualization of signals.

For the DSP control, a program was developed in C++ language executed on a microcomputer. This program allows the DSP to be verified and configured, and data acquisition to be activated and deactivated.

An online visualization resource of the acoustic emission and electrical power signals monitored by the DSP was incorporated into the computer program. This resource enables one to check if the reading of the signals is occurring correctly, thereby aiding the configuration and adjustment of the system. Using this program, it is also possible to download data corresponding to the signals and statistics stored in the memory of the DSP. These data are stored in .txt type files in ASCII format. All communication between the computer and the DSP is performed serially with the RS232 protocol.

Figure 2 shows the screen of the program executing the signal visualization resource.

LabVIEW environment, in which the data acquisition board can be configured. The signals captured by the DAQ are stored by the LabVIEW application in .bin files in binary format.

To control of DAQ system, an application was used in the All the processing of the signals from the DAQ was done in the MATLAB environment. For the

DSP system, part of the processing was performed by the device, and MATLAB was used only to calculate the mean values of the acoustic emission signal andthe electrical power. After completing all the necessary processing, the results obtained from both the systems were compared using MATLAB and Excel.

The necessary processing involves the extraction of the pass signal, filtration of the signals, and calculation of the necessary statistics.

An extraction of the pass signal involves separating from each monitored signal only the portion that corresponds to the moment when the dresser is in contact with the grinding wheel. This processing is necessary because the portion of interest for analysis is the contact between the dresser and the grinding wheel, and any data outside the pass are irrelevant.

To extract the signal, an algorithm was developed that obtains a level of reference based on samples of both the signals outside the pass and makes comparisons between new samples and the reference samples in order to determine the moment when the dresser makes contact with the grinding wheel. Because the system with DSP operates in real time, the pass extraction algorithm was executed for each new sample, and therefore only the samples referring to the pass were stored. On the other hand, for the system with DAQ operating off-line, an algorithm similar to the one implemented in the DSP was executed in MATLAB, using samples inside and outside the pass.

Signal filtration processing is necessary to eliminate noise frequencies and information irrelevant to the process. Because the acquisition frequencies used here were 10 kHz for acoustic emission and 1 kHz for electrical power, two second-order IIR digital filters were implemented, limiting the frequency of the acoustic emission signal to 30 Hz and the electrical power signal to 5 Hz, i.e., in the frequency ranges containing information relevant to the process.

After the signal extraction and filtration processing, which can be considered the preparation processing, the signals are ready for the statistical calculations, which involves calculating the values of the DPO, DPKS and the mean values of the signals. In the DSP system, the algorithm implemented in the digital signals processor performed the calculation and presented the results of the DPO and DPKS statistics, while the mean values of the signals were calculated with MATLAB software using the data obtained from the DSP.

The objective of this study was to confirm the efficiency and the possibility of implementing a DSP-based monitoring system for the dressing process, to which end a comparative analysis was made with a DAQ-based monitoring system. The algorithms used in the signals processor and in MATLAB were therefore as similar as possible and the entire processing sequence was the same, and the signals were monitored by the two systems simultaneously.

4 Results

Due to lack of space, only the results obtained with test number 5 are presented here. Similar results were obtained with the other tests.

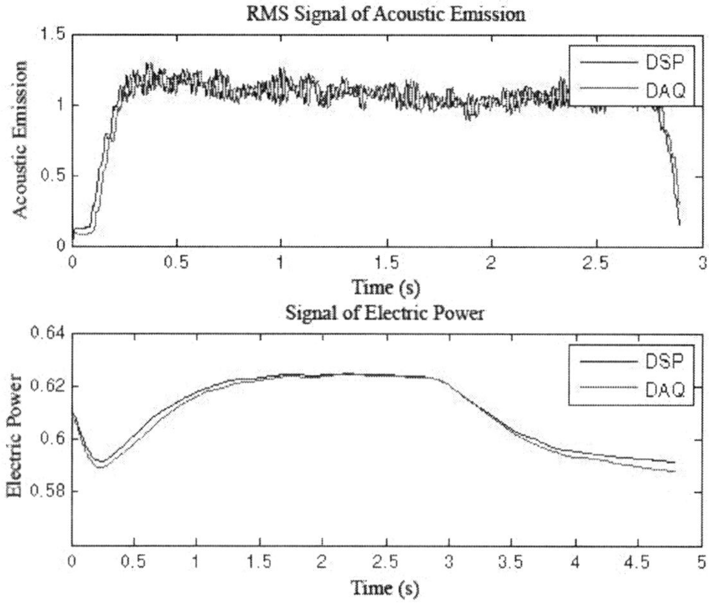

Figure 3: Curve of the fourth pass of the fifth test.

Figure 3 presents the curves of the acoustic emission and electrical power signals for pass number 4 obtained by the DSP and DAQ systems. With the help of MATLAB, the signals were superimposed to compare the quality of the data acquisition.

On these curves, the vertical scale is given in *V*k*, where *k* is a constant for the acoustic emission signal, and in *W*c*, where *c* is a constant for the electrical power signal.

An analysis of Fig. 3 reveals the similarity, in shape and behavior, of the acoustic emission signal between the DSP and DAQ systems. The predominant differences in the acoustic emission signal are the time deviation and the out of pass value that occur in the DAQ data.

The first difference is justified by the method used in extracting the pass. Because the data that are relevant for processing are only those that occur during the contact between dresser and grinding wheel, the other data must be excluded. It should be kept in mind that, for the detection and extraction of the pass, the samples of signals out of pass are used as references and the data acquisition hardware of the DSP and DAQ systems is different. Therefore, there may be differences in the sampled values, which may lead to the deviation depicted in Fig. 3.

One way to reduce this deviation is by means of a fine adjustment of the parameters of the A/D converters of both systems. However, since this difference was small and what is relevant in the processing is the behavior of the signal, the intensity of the deviation is acceptable.

In Fig. 3, note that the electrical power signal shows a greater difference, which was more marked in the first tests, whose data could not be presented here due to lack of space.

With regard to the power signal, it is important to point out the high noise load from the electrical installation of the test site. This problem prevented the use of the power pass extraction algorithm, and the acoustic emission signal therefore had to be used as time reference for the pass extraction of the electrical power signal.

Another important factor to be considered for the Power signal is the sampling rate, which was different on the two systems. The DSP used different sampling frequencies, which were 10 kHz for acoustic emission and 1 kHz for power, while the DAQ, due to a hardware characteristic, shares the same frequency between the two channels, thus precluding the use of different rates. Thus, the reading speed of power with the DAQ was 10 times that used in the DSP, generating a larger volume of points and leading to the need for a reduction by the software in order to equalize the number of points and the sampling rate. This process of reduction was performed with MATLAB before the pass extraction algorithm.

Moreover, analyzing the signals in Fig. 4c and Fig. 4d, one can see that the mean acoustic emission signal varied significantly, while the mean electrical power signal was practically constant throughout the tests. This minor variation in power is related to the dimensions of the grinding wheel and the dresser.

Because the dresser used here was a single-tip diamond with a cutting depth of 40 μm, whose width of action is between 231 and 238 μm, these dimensions are negligible when compared to the diameter of the grinding wheel. Hence, in these conditions, the passage of the dresser across the grinding wheel does not require significant additional power from the motor.

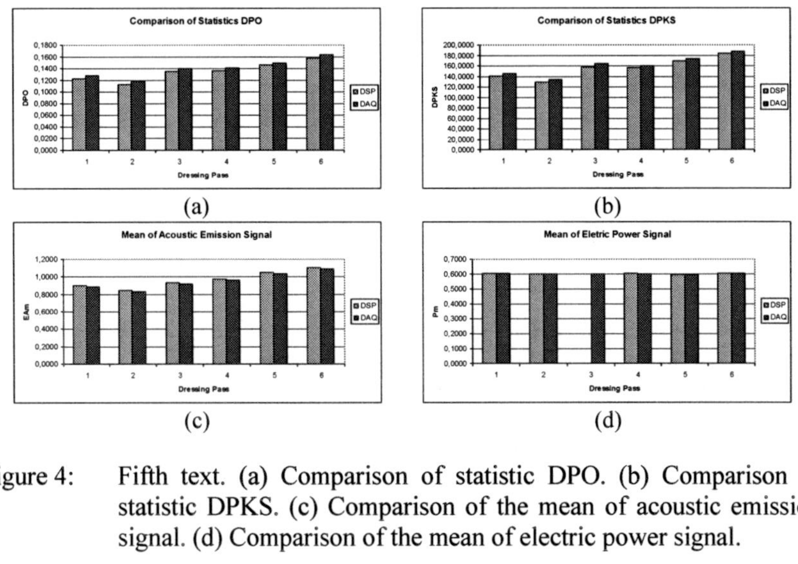

Figure 4: Fifth text. (a) Comparison of statistic DPO. (b) Comparison of statistic DPKS. (c) Comparison of the mean of acoustic emission signal. (d) Comparison of the mean of electric power signal.

Therefore, based on this consideration and analyzing all the tests, it can be proposed that the signal of interest in monitoring the dressing process is the acoustic emission signal.

An examination of Fig. 4a, Fig. 4b and Fig. 4c reveals similarities in the behavior of the mean acoustic emission signal and the DPO and DPKS parameters. One can also see that the value of these three statistics shows the same behavior in the two systems used here, with a minor error of value. This similarity was observed in all the tests.

In view of the fact that, to determine the end of the dressing process, one must observe the behavior of the DPO and DPKS statistics, which should tend to stabilize at a constant value, we can propose the substitution of the calculation of the DPO and DPKS parameters for a calculation of the mean acoustic emission signal, thus simplifying the calculations and reducing the computational effort. This proposal is possible due to the similarity of the behavior of the DPO and DPKS statistics and that of the mean acoustic emission. The behavior obtained from the statistics was as expected, i.e., their increase along with increasing passes. For the test presented in Fig. 3 and Fig 4, it can be seen that starting from the fourth pass, the value assumes a practically constant level, indicating that the process can be concluded. However, we cannot generalize and state that the grinding wheel is dressed with four passes, since the number of passes required depends on the initial conditions of the grinding wheel, the type of dresser, the dressing depth, and the table's speed.

5 Conclusion

Based on the results obtained and according to the analyses performed, it can be inferred that the use of a hardware system with a digital signal processor operating in real time mode is possible and viable from the practical standpoint, and that specific hardware can be developed to monitor the dressing process.

The main factor observed was the data acquisition capacity through the DSP without the occurrence of signal losses or distortions. After the tests and analysis of the results, it was found that the DSP hardware DSP has sufficient speed for operations in real time, without causing loss or distortion of signals.

Because the errors presented were minor, never exceeding 10%, they can be considered acceptable, considering that they can be reduced by making finer hardware adjustments.

The analysis of the results indicates that only the acoustic emission signal requires monitoring. Moreover, calculation of the DPO and DPKS parameters can be substituted for calculation of the mean effective acoustic emission signal.

Lastly, according to this study, the use of DSP on the production line and, in principle, in the dressing process, is commercially feasible and can constitute a new tool to aid this very important fabrication stage in industry, improving the product's final quality, and hence, adding value to the product.

References

[1] J.F.G. Oliveira, E.J. Silva, M. Biffi, & F. Matrai, “New architecture control system for an intelligent high speed grinder,” *Abrasives Magazine*, 2002, pp. 4–11.

[2] A. Hassui, A.E. Diniz, J.F.G. Oliveira, J. Felipe, & J.J.F. Gomes, “Experimental evaluation on grinding wheel wear through vibration and acoustic emission,” *Wear*, vol. 217, 1998, pp. 7-14.

[3] J.A. Badger & A. Torrance, “Burn awareness-Understanding the causes of grinding burn helps alleviate the problem,” *Cutting Tool Engineering Magazine*, vol. 52, 2000, p. 12.

[4] B. Karpuschewski, M. Wehmeier, & I. Inasaki, “Grinding Monitoring System Based on Power and Acoustic Emission Sensors,” *CIRP Annals - Manufacturing Technology*, vol. 49, 2000, pp. 235-240.

[5] J. Webster, W.P. Dong, & R. Lindsay, “Raw acoustic emission signal analysis of grinding process,” *CIRP Annals-Manufacturing Technology*, vol. 45, 1996, pp. 335-340.

[6] P.R. Aguiar, E.C. Bianchi, & J.F.G. Oliveira, “A method for burning detection in grinding process using acoustic emission and effective electrical power signal,” *CIRP Journal of Manufacturing Systems, Paris*, vol. 31, 2002, pp. 253-257.

[7] F.R.L. Dotto, P.R. Aguiar, E.C. Bianchi, P.J.A. Serni, & R. Thomazella, “Automatic system for thermal damage detection in manufacturing process with internet monitoring,” *Journal of the Brazilian Society of Mechanical Sciences and Engineering*, vol. 28, 2006, pp. 153-160.

[8] C.F.L.P. Hellmeister, “Monitoramento da Dresagem na Retificação através do Sinal Puro de Emissão Acústica,” Mestrado, Universidade Estadual Paulista, 2004.

[9] P.R. Aguiar, A.G. Souza, E.C. Bianchi, R.R. Leite, & F.R.L. Dotto, “Monitoring the dressing operation in the grinding process,” *International Journal of Machining and Machinability of Materials* , vol. 5, 2009, pp. 3 - 22.

[10] S. Malkin, *Grinding Technology: Theory and Applications of Machining with Abrasives*, Dearborn: SME, 1989.

[11] S. Sultana, “Effect of dressing parameters on grinding wheel wear,” Mestrado, Dalhousie University, 2004.

[12] I.D. Marinescu, W.B. Rowe, B. Dimitrov, & I. Inasaki, *Tribology of Abrasive Machining Processes*, Norwich, NY: William Andrew, 2004.

[13] S.C. Salmon, *Modern Grinding Process Technology*, New York: McGraw-Hill, 1992.

[14] J.F.G. Oliveira, D.A. Dornfeld, & B. Winter, "Dimensional characterisation of grinding wheel surface through acoustic emission," CIRP Annals, vol. 43, 1994, pp.291–294.

[15] I. Inasaki, "Monitoring of dressing and grinding process with acoustic emission signals," CIRP Annals, vol. 34, 1985, pp.277–280.

[16] J. Webster, I. Marinescu, & R. Bennett, "Acoustic emission for process control and monitoring of surface integrity during grinding," CIRP Annals, vol. 43, 1994, pp.299–304.
[17] W. Konig, & H.P. Meyen, "AE in grinding and dressing: accuracy and process reliability," Proc. of 4th Int. Grind. Conference, 1990, MR90-526.
[18] J. Kwak, & M. Ha, "Neural network approach for diagnosis of grinding operation by acoustic emission and power signals," Journal of Materials Processing Technology, vol. 147, 2004, pp.65–71.
[19] D.E. Lee, I. Hwang, C.M.O Valente, J.F.G. Oliveira, & D.A. Dornfeld, "Precision manufacturing process monitoring with acoustic emission," International Journal of Machine Tools and Manufacture, vol. 46, 2006, pp.176–188.
[20] B. Linke, "Dressing process model for vitrified bonded grinding wheels," CIRP Annals - Manufacturing Technology, vol. 57, 2008, pp. 345-348.
[21] H.T. Young, & D.J. Chen, "Online dressing of profile grinding wheels," Int. J. Adv. Manuf. Technol., vol. 27, 2006, pp. 883-888.
[22] Li Xue, F. Naghdy, & C. Cook, "Monitoring of wheel dressing operations for precision grinding," Industrial Technology, 2002. IEEE ICIT '02. 2002 IEEE International Conference on, vol. 2, 2002, pp 1296-1299.

Comparative study of the effect of PEG and Crodamide on the scratch resistance of PMMA

M. Mansha[1], C. Gauthier[1], P. Gerard[2] & R. Schirrer[1]
[1]*University of Strasbourg, Institute Charles Sadron, CNRS, France*
[2]*Arkema, GRL Lacq, France*

Abstract

Scratch durability of polymer surfaces and coatings is becoming crucial for the growing use of these materials in innovative applications, replacing former materials with more resistant surfaces. A variety of physico-chemical processes such as annealing and a range of ion plantation techniques have been proved very valuable for the renovation of polymeric surfaces but their applications are restricted to some certain polymers due to a number of disadvantages. In particular, several of them can not be applied to translucent polymers like PMMA due to their darkening outcome. Enhancement of the scratch resistance must be investigated for the most part as a consequence due to decreasing the friction coefficient. The recovery increases if the tip is smooth or if the local friction coefficient is low. Discovery of appropriate plasticizers for different polymers to decrease their surface friction is yet to be made in order to improve their scratch resistance.

The effect of PEG and crodamide on both surface properties and bulk mechanical properties of PMMA was studied over a wide range of temperature (ranging from -40 to +85°C). Our experimental results show that a decrease in friction coefficient is possible by the introduction of appropriate plasticizer without having a significant effect on its bulk behaviour and this decrease in friction depends upon the nature and the content of plasticizer. Moreover, fatty acid amides have been proved more effective in decreasing friction than PEG. The friction decreases directly with the percentage of crodamide. The values of the rear contact angle and in-situ photographs during scratching confirm that the decrease in friction is associated with the decrease of yielding of the contact between the tip and the polymer surface.

Keywords: scratch resistance, friction coefficient, polymer, PMMA, crodamide, PEG, bulk properties.

WIT Transactions on Engineering Sciences, Vol 66, © 2010 WIT Press
www.witpress.com, ISSN 1743-3533 (on-line)
doi:10.2495/TD100151

1 Introduction

Polymers are becoming more and more attractive for a growing number of industries. Their low cost, ease and simplicity of manufacture and processing, low weight as well as the attractive look they furnish to any surface have amplified their use [1–5]. Abrasion and scratching are by and large the cause of the loss of their optical and tribological performances [6–9]. The surfaces of polymeric materials are predominantly sensitive to scratches, which obstruct their use for high performance optical devices. In other state of affairs, these materials are used for moving mechanical parts, where friction and wear properties are important [10]. Scratches on plastic surfaces seriously trim down the surface aesthetics of the polymer products and also lead to stress concentration, making products prone to premature failure. That is why there is a strong call for to get better scratch resistance of polymers for automotive and other applications that insist high surface excellence [3, 11].

It is normally acknowledged that surface damage can be divided into two types; mar and scratch. A mar is a spot due to a sliding body that is excessively low to be ostensible by the casual human eyes alone but however does turn into noticeable when present in huge quantities. The classic damage found on paint coats and dashboard surfaces smashed by tiny objects such as rough stones, sticks, keys, etc are the examples. A scratch is a mark that makes perceptible grooves and/or surface damage; this is the distinctive damage mode for surfaces that endure profound moving loads by swivels, ball bearings, etc. [2, 12]. The asperities on the surface of a hard material, if they come in contact, can easily generate scratches on the polymer surface. The infrequent occurrence of hard debris or third-body, as they are referred to, can also be a cause of scratches on the polymer surfaces [13]. Scratches may be of any size and length, and the perceiving, or valuation, of the scratch is generally linked with its look [11].

Scratch method offers an appropriate and trustworthy way to study the mechanical properties of organic polymers under different contact conditions [6, 10, 11, 13–16]. A number of researchers have used scratch test to evaluate resistance to scratch for different polymers. Owing to the large elastic recovery of polymers and their complex rheological behaviour, the analysis and the elucidation of this test, however, is far more complicated than for metals [14]. Chief variables that can modify the damage response of polymers during scratching are strain (defined by the angle of attack), scratching velocity, normal load (that determines the depth of scratch), temperature or thermal effects, state of lubrication and the stiffness property of the scratching device. In order to slim down various effects, Briscoe et al. [11] have proposed the idea of using scratching maps while presenting scratching data. Although qualitative in nature, scratching maps provide an outstanding way of identifying some broad deformation behaviour of polymers such as ductile, brittle or visco-plastic for a broad array of experimental parameters [13, 15].

Jardret et al. have revealed three kinds of deformation during the experiments: plastic or stable deformation, surface elastic deflection in the region of the contact zone, and the elastic recovery of the scratch groove. Measurement of the scratch width permits the assessment of the contact depth and therefore of the contact pressure between the indenter and the material's surface [16]. Sue and co-workers have concentrated to apply the scratch test for evaluating the surface scratch resistance of different polymers. They noticed that the scratch process is a function of the basic material properties such as the elastic modulus, the yield strength and the tensile strength as well as material surface characteristics such as the coefficient of friction [2, 3, 15].

The contact area between a moving tip and a polymer surface has a front side and a rear side and its shape changes with the strain [12, 17, 18]. A contact area which is totally plastically deformed is called for simplicity a 'viscoplastic contact'. If the contact area is not entirely plastically deformed, it is called a 'viscoelastic–plastic contact'. The shape of the contact is simply described by the rear angle or by the ratio of the rear length to the front length. The rear contact angle ω is due to the elastic recovery of the polymer and depends on the plastic deformation around the contact. Figure 1 gives a schematic illustration of the evolution of the rear contact angle where as Figure 2 represents contact angle as a function of the mean contact strain ε. ε_e and ε_p represent the contact strains at the end of the elastic contact area and the commencement of the plastic contact area, respectively. In the case of a viscoelastic contact, the rear angle ω is equal to $\pi/2$ for the elastic deformation. If the material shows strain hardening, the contact strain decreases and ω increases. As the strain rate and temperature vary, even under constant loading, the contact area may also vary significantly and the strain in the surrounding area of the contact may change from viscoelastic to viscoplastic [17, 18]. The material underneath a moving tip may demonstrate diverse types of behaviour: elastic, elastic–plastic where elastic and plastic strains are present in the contact area, or fully plastic. As is accustomed in polymers, the material behaviour is time and temperature dependent and may display strain hardening [18, 19].

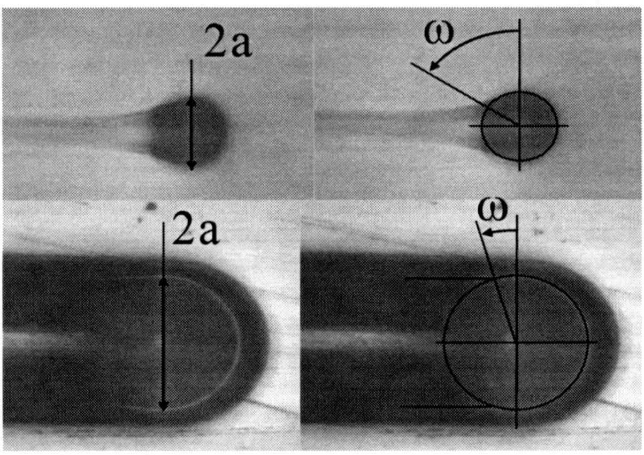

Figure 1: Photograph of the contact area during scratching. This area has a rear part defined by the angle ω.

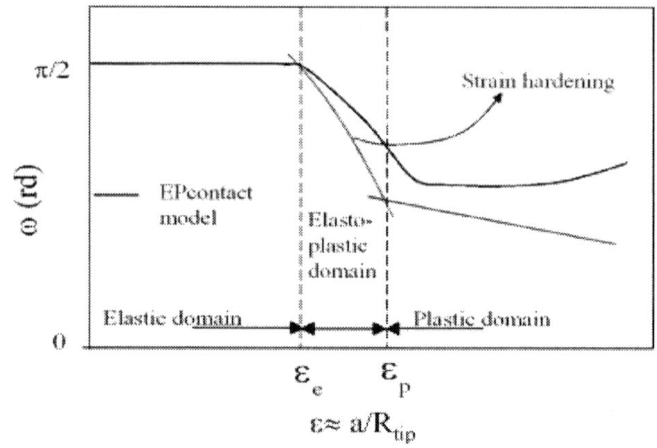

Figure 2: Rear contact angle as a function of the mean contact strain. This angle decreases in the case of a viscoelastic–plastic or viscoplastic contact and increases if the material shows strain hardening [17, 18].

Another important property that is universally studied by means of a scratch test is the overall friction coefficient defined as the ratio of the tangential force to the normal force. It is normally believed to take place from two mechanisms: adhesion and ploughing. It has been found that the overall friction coefficient in a scratching test is influenced by many factors such as material type, sliding speed, lubricating conditions, and the indenter geometry. The experimental results by Briscoe et al. indicated that the ratio of adhesive interfacial force to the total scratch force tended to unity when the cone angle of the indenter was increased in single-pass scratch tests. This may also imply that the ploughing effect decreases with the increasing cone angle of the indenter [20]. The polymer true friction coefficient displays temperature and sliding speed dependency, which may be recognized due to the surface thermodynamics. It was shown by Lafeye et al. that the local friction depends on the level of strain in the polymer at the contact interface [18, 19].

Gauthier et al. observed the transition from viscoplastic scratching to elastic sliding and found that the temperature, strain and strain rate are the important parameters controlling the nature of scratching on viscoelastic viscoplastic materials like polymers [10]. They proved for the macroscopic scale that the mechanical performance of the contact is tailored by the decrease in the friction coefficient. They also presented an analysis of the elastic recovery of a scratch after contact with a tip. They concluded that the recovery increases if the tip is smooth or if the local friction coefficient is low. So the upgrading of scratch resistance should be investigated principally as an outcome caused by decreasing the friction coefficient [17]. Exploration of the reduction of friction through use of different plasticizers is an area that has an immense potential of research. Plasticizers are normally used to perk up flexibility, trim down cost and improve the processability of polymers. Preferably, plasticizers would be low in price, compatible with the host polymer and be processed without difficulty. For extending the life span of plasticized materials, the additive ought to have a low volatility and should not seep from the host. The plasticizer should also give reproducible control over material properties, by adjusting the chemical structure and the quantity of the plasticizer used [21].

2 Experimental details

2.1 Materials and methodology

2.1.1 Materials

The PMMA studied was a standard grade PMMA (V825), having molecular weight about 100K. It was provided by Arkema. The effect of plasticizers was studied by testing the samples containing 0, 1, 3, and 5 % of either of the both; i.e. PEG and Crodamide. PEG is generally recognised due to its low cost and antiblock properties [22]. Owing to compatibility with most of polymers, fatty acid amides are being used world widely to overcome many of the slip, blocking and mould-release problems faced by the polymer industry [23]. Crodamide, like other fatty acid amides, has superior thermal stability which permits its use at relatively higher process temperatures where use of other conventional additives is not recommended [24, 25]. PEG [(C_2H_4O) nH2O] employed had Mol. Wt. 3350 and was purchased from Sigma-Aldrich where as Crodamide used was Crodamide ER (or refined erucamide) and was purchased from Croda Chemicals.

2.1.2 Scratching apparatus and the indenter

The experimental data was obtained with a scratch apparatus fitted with a built-in microscope, which allows in situ examination of the contact area and the groove left on the surface (which is practicable because of the lucidity of the tested polymers). The scratch apparatus, described in detail by Gauthier and Schirrer, is known as the microvisioscratch apparatus. It is based on a commercial servomechanism having a small translucent environmental compartment, which contains the sample and the moving tip (Figure 3). Scratch test may be performed over a large array of speeds (1 μ m/s to 15 mm/s) and in a temperature range covering both α and β transitions of all-purpose polymers (– 70°C to +120°C). The normal load applied to the tip can be varied from 0.05 N to 35 N. Control of the moving tip and the recording of the load, the speed of the sliding tip and the temperature are computer driven. The tip starts at the lowest value of the velocity (or at the lowest value of normal load in the alternate mode) and accelerates stepwise up to the highest value. At each step, it moves over a distance of at least 1 mm so as to get a groove that can be easily measured and photographed in-sit. The input parameters are the normal load, tip geometry, temperature and the sliding speed, whereas the output parameters are the tangential force, groove geometry and the true contact area.

All samples were scratched by means of a conical scratcher (tip radius =0.116 mm and apex angle=60°) with a varying load from 0.05N to about 5 N

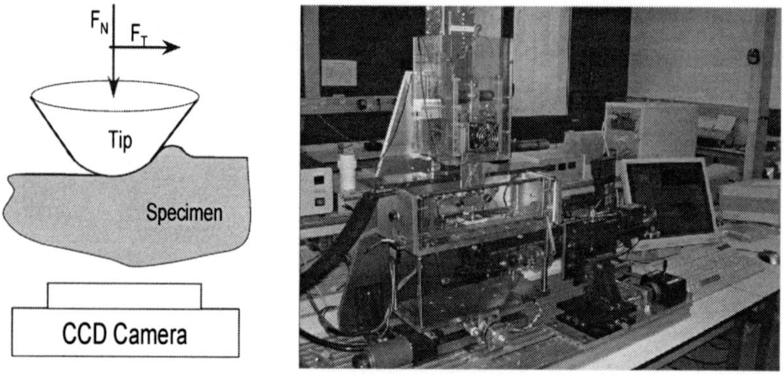

Figure 3: A schematic diagram (on the left) and the photograph (on the right) of the friction measuring apparatus.

depending upon the test temperature (ranging from -40°C to 85°C). The scratcher was made of diamond, and the sliding speed was maintained at about 0.03 mm/s. A typical procedure was used to perform the friction tests [19]. Sliding and scratching tests are characteristically carried out at mounting normal load at a constant speed for a given temperature. At every loading step, the true contact area is photographed in situ. This image helps one to find out the shape and width of the contact area [26].

2.1.3 Measurement of bulk properties

All the specimens with different level of plasticizer content were tested using Instron 4015 Universal Testing Machine with capacity of 10 kN and ram velocity ranging from 1.0 μm/sec to 10.0 mm/sec. The main objectives of the tests are to obtain the values of tensile strength and strain, shear stress and strain at break, elongation at breakage and the relation of these values versus plasticizer contents for PMMA.

3 Results and discussion

3.1 Surface properties

3.1.1 Coefficient of friction

As discussed in section 1, the apparent friction coefficient [μ_{app}] is the proportion between the tangential force and the normal load applied to the moving body in contact with the surface of a material. This coefficient includes a so-called ''true local friction'' [μ_{loc}] - also named as "adhesive friction" [μ_{ad}] in case of perfectly smooth surface and a ''geometrical friction'' which is the ploughing effect. The latter may be decomposed into two terms; one term due to the viscoelasticity and the other due to the plasticity. Hence the friction may be written as [12, 18, 19].

$$\mu_{app} = \mu_{ad} + \mu_{visco} + \mu_{plast} \tag{1}$$

When a slip agent (a plasticizer being used for reducing the friction) is mixed in a polymer melt, it is absorbed in polymer region. It remains there as long as polymer remains in melt form. When the polymer solidifies at lower temperature, the additive becomes incompatible with the polymer due to difference in surface energy and start migration towards surface. Fatty acid amides are characterised for forming a thin film over the surface of the finished product giving an excellent slip effect [22]. Our experimental results indicate that the crodamide is more effective than PEG in reducing friction coefficient of PMMA (Figures 4 and 5).

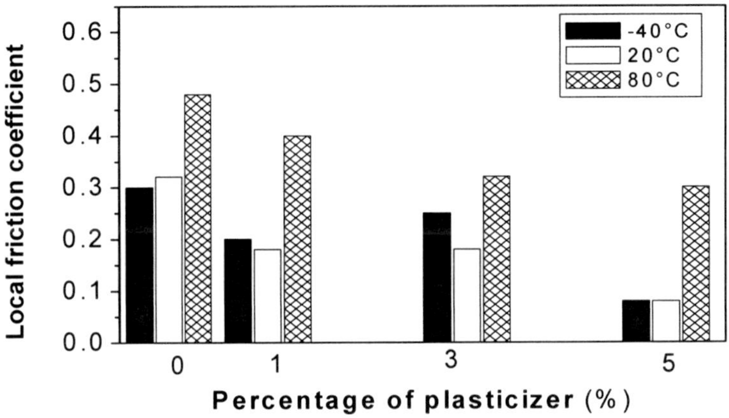

Figure 4: Effect of crodamide on local friction coefficient of PMMA at -40°C, 20°C and 80°C (sliding speed=0.03 mm/sec, sliding tip radius=116μm, normal load=1N).

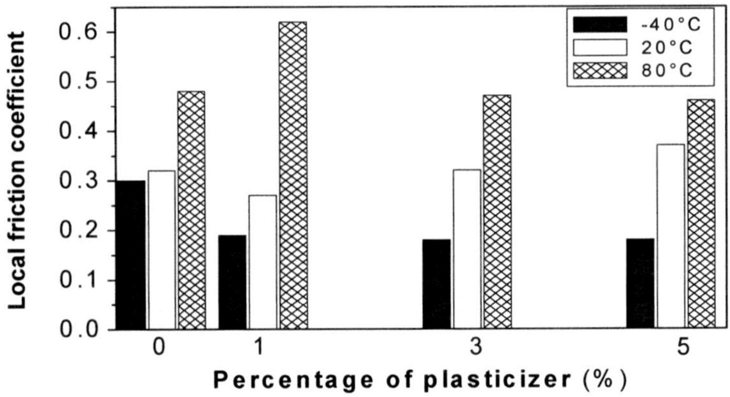

Figure 5: Effect of PEG on local friction coefficient of PMMA at -40°C, 20°C and 80°C (sliding speed=0.03 mm/sec, sliding tip radius=116μm, normal load=1N).

The decrease in friction coefficient increases with increasing crodamide content. On the other hand, in case of PEG, the coefficient of friction decreases initially on the introduction of PEG (up to 1%) but, on further addition of PEG, no additional reduction in the friction coefficient is observed. Moreover, at higher temperature, no any significant effect of addition of PEG on the reduction of the friction coefficient of PMMA is noticed. Due to higher molecular weight of PEG, it is difficult to be migrated to the surface and to form a film lowering the friction. The values of bulk properties like young modulus of the two cases (as shown in next section) also support this conclusion.

3.1.2 Contact angle

As discussed in section 1, the contact angle (omega) is the indicator of the nature of the contact between the sliding tip and the polymer surface. Our experiments indicate that the introduction of crodamide in PMMA modifies the nature of the contact and enhances the visco-elasticity in the contact. If we see, for example, the values of contact angle for the samples containing different percentages of crodamide at room temperature (Figure 6), we come to know that at same value of contact strain (0.05), the value of omega is changed significantly by the introduction of crodamide. It is about 0.4 for pure PMMA (having higher value of local friction coefficient) that increases to 0.9 for crodamide –plasticized PMMA (having lower value of local friction coefficient). So it can be concluded that the decrease in friction coefficient is associated with the increased value of the contact angle and ultimately with the nature of the contact. In situ photographs of the contact demonstrate that the contact nature is plastic for pure

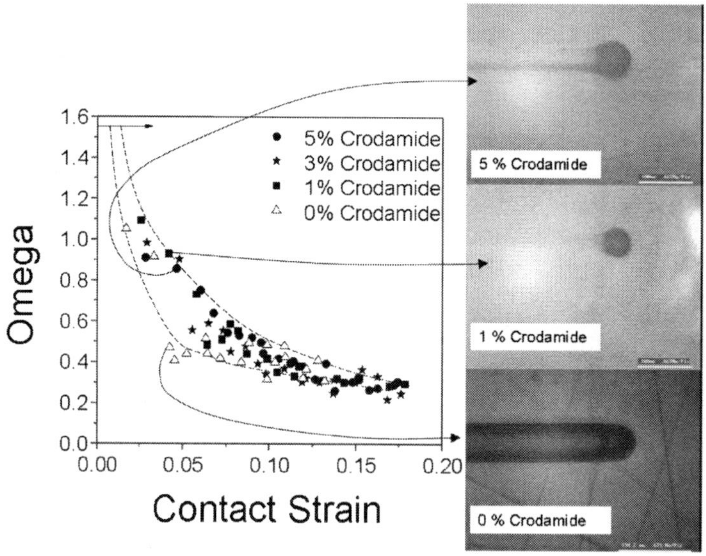

Figure 6: Variation of omega with contact strain and the percentage of Crodamide for PMMA at room temperature (sliding/scratching speed =0.03mm/sec, tip radius =116μm).

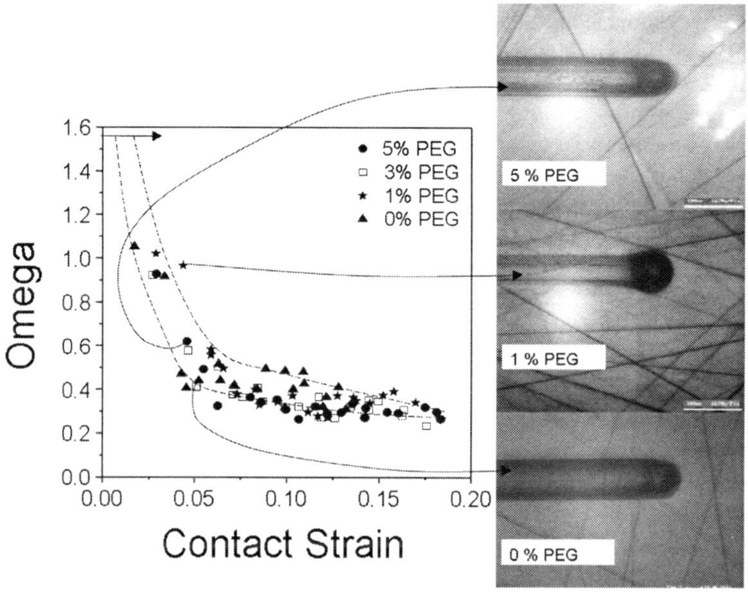

Figure 7: Variation of omega with contact strain and the percentage of PEG for PMMA at room temperature (sliding/scratching speed =0.03mm/sec, tip radius =116μm).

PMMA at contact strain about 0.05. On the other hand, it is viscoelastic for crodamide-plasticized samples at the same value of contact strain (Figure 6).

It is interesting to note that at the same conditions of temperature and contact strain, the nature of contact for PEG plasticized samples of PMMA is similar to that of pure PMMA, i.e. almost plastic (Figure 7). The values of the contact angle for PEG plasticized samples also remain almost same as that of pure PMMA for a given value of contact strain. The little decrease in friction for the samples containing 1% PEG is also reflected in the nature of contact as is clear from the in-situ photograph and the value of contact angle is also increased a little bit correspondingly. However, at a given condition of temperature and normal load, the contact is purely elastic for all samples at very low contact strains and is purely plastic at very high values of contact strain (Figures 6 and 7).

3.2 Bulk properties

Elastic Modulus and Yields Stress are the bulk material properties affecting the scratch process.

3.2.1 Elastic modulus

Elastic modulus displays a principal part in the scratch process, particularly for the duration of the indentation stage. According to the scratch model, the Young's modulus, scratching tip geometry, and Poisson's ratio directly decide the range of the contact zone during scratch. For polymers with lower modulus, the contact zone dimension increases a lot with decreasing modulus. It is fascinating to note that for polymers with high Young's modulus, the consequence of the change in the Young's modulus on the contact zone size becomes immaterial.

The penetration depth is more responsive to the modulus when it is inferior to 1 GPa. The maximum tensile stress is on the surface immediately at the rear of the scratcher during scratch. It is clear that polymers with higher moduli demonstrate a higher maximum tensile stress on the surface. This may result in the development of surface cracks, crazes, cavitations, and debonding between phases of multiphase polymers. As mentioned above, the Young's modulus plays an important role in generating scratch damage. A certain level of stiffness in polymers can assist to maintain a low scratch depth if the polymer is inclined to plastic flow scratch damage. In contrast, polymers with elevated rigidity induce high stress concentration during scratch, which could prop up the creation of surface cracks, crazes, cavitations, and debonding. In this case, it is preferable to lower the modulus or enhance the ductility of the polymer [3].

As the percentage content of plasticizer in all cases was not so high, their consequence on the mechanical properties is not up to significant value (Figure 8). Both crodamide and PEG have no remarkable effect on major bulk properties like Young’s modulus. However, PEG disturbs these properties comparatively more than does crodamide. This difference of bulk behaviour can be attributed to the enhanced tendency of migration of crodamide from bulk to the surface of the material as is discussed in previous section. Due to higher molecular weight of PEG, it migrates to the surface a little and mainly remains inside the bulk of material and hence disturbs the bulk behaviour more pronouncedly as compared to the crodamide.

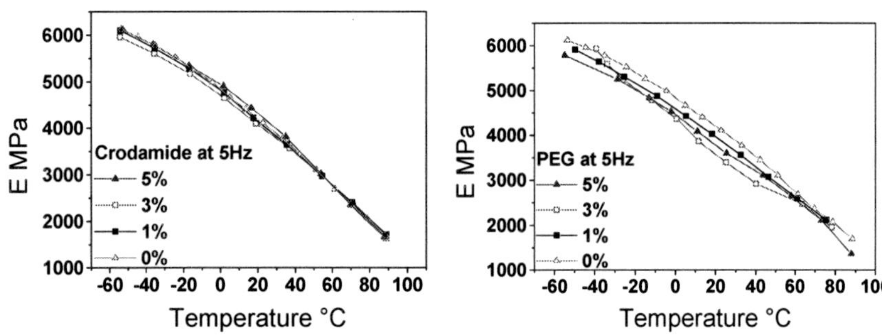

Figure 8: Effect of crodamide and PEG on young’s modulus of PMMA.

3.2.2 Yield stress

The yield stress plays a prime role to determine the upper boundary of the mean contact pressure and the size of the plastic zone. Polymers having an elevated yield stress are subjected to smaller plastic flow scratch damage. On the other hand, cracks are generally generated on the scratch surface if the utmost tensile stress on the surface exceeds the tensile strength and if the tensile strength is lesser than the yield stress of the polymer. Therefore, the tensile strength of the polymer plays a key role in determining the growth of either the plastic flow scratch pattern or the brittle fracture scratch pattern on the scratch surface [3].

The effect of both crodamide and PEG on nominal stress and nominal strain at break is not very clear. Generally speaking, both increase these values a little bit for PMMA and this is more pronounced in case of PEG (Figure 9). However, at low temperature, crodamide first decreases the nominal stress and strain at break of PMMA but further increase in its content again increase the nominal stress and strain values of PMMA. Similarly the effect of both Crodamide and PEG on shear stress and shear strain at breakage is not very clear (Figure 10). Shear stress and strain at breakage is affected a little by both the crodamide and PEG and this effect is more significant in case of crodamide. However, at higher temperature, this effect is more pronounced for PEG.

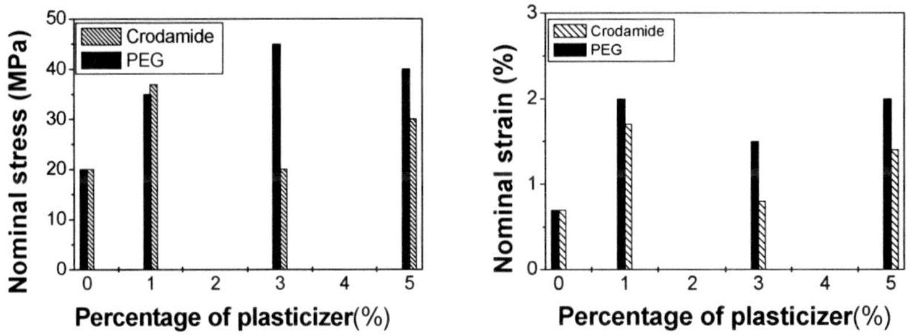

Figure 9: Effect of crodamide and PEG on nominal stress and strain at break of PMMA at 40°C (traction speed=10^{-4}/sec).

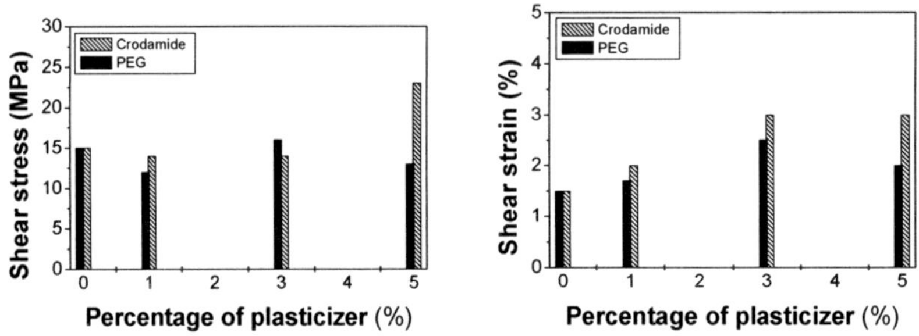

Figure 10: Effect of crodamide and PEG on shear stress and strain at break of. PMMA at 40°C (traction speed=10^{-4}/sec).

4 Conclusions and prospective

The following conclusions can be drawn:

1. The coefficient of friction of PMMA can be decreased by the use of appropriate plasticizer.

2. The crodamide is more effective than PEG for reducing the friction coefficient of PMMA and the decrease in friction coefficient is proportional to the crodamide content.
3. The decrease in friction coefficient of PMMA by crodamide is not at the cost of bulk properties. They remain almost unaffected.
4. The decrease in friction is associated with the nature of the contact between the tip and the polymer surface.
5. The use of crodamide is also favourable as it improves the bulk properties a little bit along with improving the surface properties in contrast to PEG, which reduces the bulk properties a little bit.
6. Generally speaking, nominal stress and nominal strain at break are increased by both crodamide and PEG but at low temperatures it is decreased a little bit by crodamide.
7. Similarly, shear stress and shear strain at break are increased by both crodamide and PEG.

However, further work is indispensable to enumerate the well balanced and optimised study surrounded by the upgrading of the surface behaviour (like reduction in friction coefficient) and the subsequent change in bulk properties.

Acknowledgements

The financial support provided by the Higher Education Commission, Pakistan is gratefully acknowledged. The authors would thank and express their sincere appreciation to Mr. Favier Damien for his help during the experimental work.

References

[1] Hamouda, A.M.S., *The influence of humidity on the deformation and fracture behaviour of PMMA.* Journal of Materials Processing Technology, **124**(1-2): p. 238-243. 2002.

[2] Wong, M., et al., *A new test methodology for evaluating scratch resistance of polymers.* Wear, **256**(11-12): p. 1214-1227. 2004.

[3] Xiang, C., et al., *Scratch behavior and material property relationship in polymers.* Journal of Polymer Science Part B: Polymer Physics, **39**(1): p. 47-59. 2001.

[4] Gauthier, C. and R. Schirrer, *Time and temperature dependence of the scratch properties of poly(methylmethacrylate) surfaces.* Journal of Materials Science, **35**(9): p. 2121-2130. 2000.

[5] Pelletier, H., et al., *Viscoelastic and elastic-plastic behaviors of amorphous polymeric surfaces during scratch.* Tribology International, **41**(11): p. 975-984. 2008.

[6] Jardret, V. and P. Morel, *Viscoelastic effects on the scratch resistance of polymers: relationship between mechanical properties and scratch properties at various temperatures.* Progress in Organic Coatings, **48**(2-4): p. 322-331. 2003.

[7] Adams, M.J., et al., *An experimental study of the nano-scratch behaviour of poly(methyl methacrylate).* Wear, **251**(1-12): p. 1579-1583. 2001.

[8] Tanglumlert, W., et al., *Hard-coating materials for poly(methyl methacrylate) from glycidoxypropyltrimethoxysilane-modified silatrane via a sol-gel process.* Surface and Coatings Technology, **200**(8): p. 2784-2790. 2006.

[9] Kopesky, E.T., G.H. McKinley, and R.E. Cohen, *Toughened poly(methyl methacrylate) nanocomposites by incorporating polyhedral oligomeric silsesquioxanes.* Polymer, **47**(1): p. 299-309. 2006.

[10] Gauthier, C., S. Lafaye, and R. Schirrer, *Elastic recovery of a scratch in a polymeric surface: experiments and analysis.* Tribology International, **34**(7): p. 469-479. 2001.

[11] Briscoe, B.J., et al., *Scratching maps for polymers.* Wear, **200**(1-2): p. 137-147. 1996.

[12] Sujeet K Sinha, B.J.B. *Polymer Tribology*. 2009.

[13] Sinha, S.K. and D.B.J. Lim, *Effects of normal load on single-pass scratching of polymer surfaces.* Wear, **260**(7-8): p. 751-765. 2006.

[14] Felder, E. and J.-L. Bucaille, *Mechanical analysis of the scratching of metals and polymers with conical indenters at moderate and large strains.* Tribology International, **39**(2): p. 70-87. 2006.

[15] Briscoe, B.J. and S.K. Sinha, *Scratch Resistance and Localised Damage Characteristics of Polymer Surfaces - a Review.* Materialwissenschaft und Werkstofftechnik, **34**(10-11): p. 989-1002. 2003.

[16] Jardret, V., et al., *Understanding and quantification of elastic and plastic deformation during a scratch test.* Wear, **218**(1): p. 8-14. 1998.

[17] Gauthier, C., et al., *Scratching of a coated polymer and mechanical analysis of a scratch resistance solution.* Tribology International, **39**(2): p. 88-98. 2006.

[18] Lafaye, S., C. Gauthier, and R. Schirrer, *A surface flow line model of a scratching tip: apparent and true local friction coefficients.* Tribology International, **38**(2): p. 113-127. 2005.

[19] Lafaye, S., C. Gauthier, and R. Schirrer, *Analysis of the apparent friction of polymeric surfaces.* Journal of Materials Science, **41**(19): p. 6441-6452. 2006.

[20] Subhash, G. and W. Zhang, *Investigation of the overall friction coefficient in single-pass scratch test.* Wear, **252**(1-2): p. 123-134. 2002.

[21] Scott, M.P., M. Rahman, and C.S. Brazel, *Application of ionic liquids as low-volatility plasticizers for PMMA.* European Polymer Journal, **39**(10): p. 1947-1953. 2003.

[22] Wypych, G., ed. *Handbook of Antiblocking, Release and Slip Additives;* 2005.

[23] http://www.atskimya.com/msds/spect59.pdf.

[24] http://www.thefreelibrary.com/Lubricants+and+processing+aids.a0189051619.

[25] Markarian, J., *Slip and antiblock additives: surface medication for film and sheet.* Plastics, Additives and Compounding, **9**(6): p. 32-35.2007

[26] Lafaye, S., C. Gauthier, and R. Schirrer, *Analyzing friction and scratch tests without in situ observation.* Wear, **265**(5-6): p. 664-673. 2008.

Plasma Electrolytic Oxidation (PEO) coatings on a zirconium alloy for improved wear and corrosion resistance

Y. Chen, X. Nie & D. O. Northwood
Department of Mechanical, Automotive and Materials Engineering, University of Windsor, Canada

Abstract

Zr oxide coatings were deposited using plasma electrolytic oxidation (PEO) on Zr-2.5wt%Nb (Zr-2.5Nb) alloy, which is currently used for pressure tubes in the CANDU nuclear reactor. The effects of two PEO processing factors, electrolyte concentration and current density, on the microstructure and properties of the coatings were systematically investigated. The coating morphology and chemical composition were determined using scanning electron microscopy (SEM) and energy dispersive X-ray analysis (EDX). Potentiodynamic polarization corrosion testing was used to examine the corrosion resistance of the coatings in a dilute aqueous, LiOH solution. Pin-on-disc wear tests were performed under dry and lubricated sliding conditions. SEM was also used to characterize the wear traces on the coated and uncoated surfaces. A 30-day autoclave experiment was carried out to study the corrosion performance of the Zr-2.5Nb substrate, PEO coatings and a commercial autoclaved black oxide coating. Based on the results of this research, recommendations are made as to processing parameters for the production of oxide coatings with improved wear and corrosion resistance.
Keywords: Zr-2.5Nb, PEO, commercial black oxide coating, corrosion and wear resistance, autoclave.

1 Introduction

Zr-2.5wt%Nb (hereafter, referred to as Zr-2.5Nb) has been used in CANDU-PHW type nuclear reactors for the manufacture of pressure tubes (PT) because of its low neutron absorption cross-section, high corrosion resistance, high strength and creep resistance [1–4]. The inside surfaces of the PTs are in contact with

WIT Transactions on Engineering Sciences, Vol 66, © 2010 WIT Press
www.witpress.com, ISSN 1743-3533 (on-line)
doi:10.2495/TD100161

lithiated heavy water (D_2O) coolant at 250 to 300 °C and as a result of a corrosion reaction, deuterium (hydrogen) is formed [2]. Between 5 and 10% of this deuterium may be absorbed by Zr-2.5Nb during normal operation which could degrade the mechanical properties. The cold-worked Zr-2.5Nb alloy PTs are autoclaved in steam at 400°C for 24hrs to produce a black adherent oxide film of 1 to 2μm in thickness which has proven to be an excellent hydrogen permeation barrier [2]. CANDU-PHW reactors utilize on-power refuelling and there is wear between the fuel bundle wear-pads and the inside surface of the PT which can lead to localized material degradation. Recent trends to higher fuel burn-up, extended recycle, high PH operation and higher operating temperatures, have led to the need of improve the corrosion and wear resistance of Zr-2.5Nb.

Plasma electrolytic oxidation (PEO) is a technique by which a oxide or other ceramic coating can be grown in-situ on Al, Ti, Mg and many other light metals [5–11]. In PEO technology, sparks or arc plasma microdischarges in an aqueous solution are used to ionize gaseous media from the solution with the result that complex compounds are synthesized on the metal surface by the plasma-chemical interactions. It was used, first of all, for the surface hardening and enhancement of the corrosion resistance of Al [5–9] and, more recently, Mg and Ti alloys [10–12]. Thus far, it has had limited application to Zr alloys [12, 13].

2 Experimental details

Square coupons with dimensions of $15\times13\times4mm^3$ were cut from commercial Zr-2.5Nb PT with black oxide coating on both surfaces (fig 1). The coupons were carefully polished flat to obtain uniform a surface roughness of R_a=0.1±0.02μm. Other coupons with the black oxide coating retained were cut to a similar geometric size for comparison.

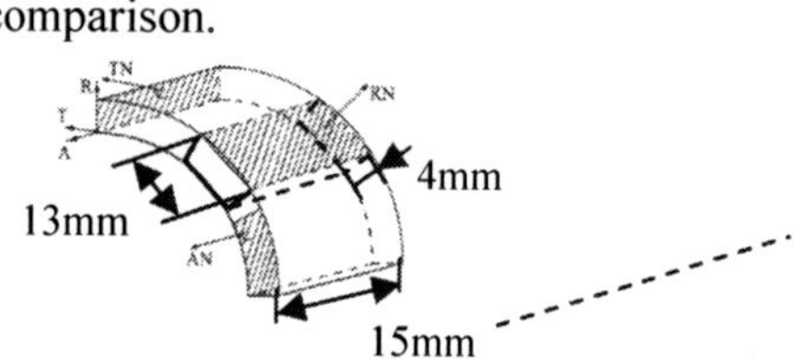

Figure 1: Schematic diagram showing the sample coupon cutting method.

A PEO technique was used to deposit ceramic coatings on Zr-2.5Nb substrates using pulsed DC power with a frequency of 2000Hz supply in a silicate: KOH electrolyte. Two different Na_2SiO_3: KOH solutions with concentration ratios of 8g/L: 0.8 g/L and 8g/L: 8 g/L were used. Different treatment times and current densities were used (table 1). Near constant thickness (5μm) coatings were produced on all coupons.

Potentiodynamic polarization tests were carried out using a Bio-Logic SP-150 potentiostat/galvanostat for uncoated, PEO-coated and inside surface of black oxide samples at room temperature (20 ºC) in a 4.8 wt.% (0.2mol/L or 1400ppm Li^+) LiOH solution, which is very corrosive to Zr alloys [14, 15]. The exposure area of the samples was 1 cm2. Scans were conducted from -0.5V to 2V.

Table 1: Treatment parameters for PEO coatings.

Composition (Na_2SiO_3:KOH)	Sample No.	Current density(A/cm^2)	Treatment time (min)
Group A Electrolyte I (Ph=11-12) (8g/L: 0.8 g/L)	S1	0.05	4
	S2	0.1	2
	S3	0.25	0.8
Group B Electrolyte II (Ph=13-14) (8g/L: 8g/L)	S4	0.05	4
	S5	0.1	2
	S6	0.25	0.8

An accelerated corrosion test involving 30-day autoclave exposure to high temperature (300°C), high pressure (10MPa) and 0.05mol/L LiOH solution was carried out to simulate the service conditions of PTs [16]. The test was interrupted at 2, 5, 10, 15, 30 days to refresh the LiOH solution and make measurements. In each period, the samples were removed from the autoclave, dried and weighed. Some samples of selected exposure times were then examined by potentiodynamic polarization tests. The others were then put back into the autoclave for another cycle.

The tribological properties were tested by a pin-on-disc tribometer at room temperature. Two modes were used: rotating mode (sliding speed: 0.1 m/s) for flat samples (both substrate and the PEO samples) and reciprocating mode (sliding speed: 0.08 m/s) for the curved samples (black oxide coating). All the PEO coatings were first tested at 2N normal load and 1000m sliding distance under dry conditions. The uncoated substrate was also tested at the same conditions and load but only to a 50m sliding distance. Surface roughness was measured across the wear trace to study its width and depth. One of the PEO coatings (S1), which showed good wear resistance, was chosen for reciprocating wear tests to compare it with the black oxide coating. In the reciprocating wear test, both dry and lubricated conditions at 2 and 5N load respectively were used. The sliding distance was 1000m but the test was stopped at coating failure.

JEOL Scanning Electron Microscope at an operating voltage of 18keV with EDX were employed to study coating thickness, surface morphology and chemical composition.

3 Results and discussion

3.1 Characterization of PEO coatings

As seen in table 1, six sets of PEO coatings were made. SEM examination was made of both the coating surface and cross-section through the sample thickness.

SEM micrographs of selected PEO coatings (S1, S2 and S5) and the commercial black oxide coating are shown in fig 2 for the surface morphology and fig 3 for the cross-sections. Sample S1, which was fabricated at a current density of $0.05A/cm^2$, exhibited a relatively dense, continuous oxide film, see figs 2(a) and 3(a). Samples S2 and S5, which were fabricated at a higher current density (0.1A/cm^2) exhibited significant porosity (figs 2(b) and (c) and figs 3(b) and (c)). At higher current densities, the discharges are more intense and its pores act as discharge channels. The black oxide coating is about 1μm thick and also contains porosity: The porosity is finer than in PEO samples S2 and S5.

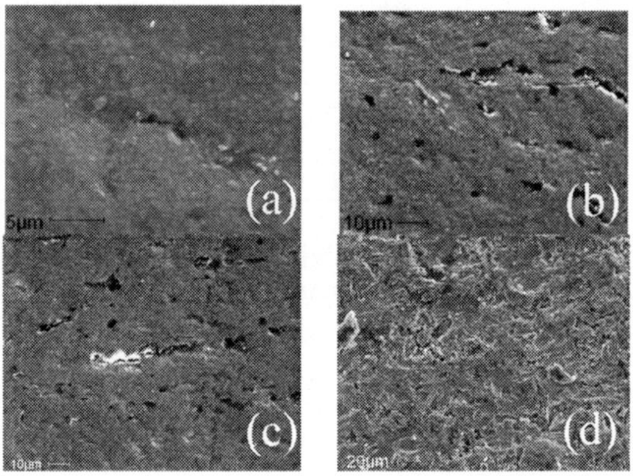

Figure 2: SEM micrographs on the morphology of (a) and (b) S1 and S2 in electrolyte I; (c) S5 in electrolyte II; (d) Black oxide coating.

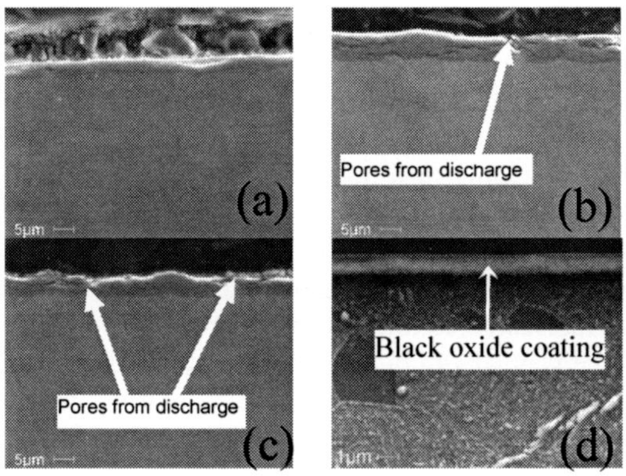

Figure 3: Cross-sectional SEM micrographs at different magnifications of the PEO coatings. (a) and (b) S1 and S2 in electrolyte I; (c) S5 in electrolyte II; (d) Black oxide coating.

3.2 Corrosion tests

3.2.1 Potentiodynamic polarization tests

Fig 4 shows the polarization curves for the Zr-2.5Nb substrate both without and with the black oxide coating and PEO coated samples. Values for the corrosion potential (E_{corr}) and corrosion current (i_{corr}) and polarization resistance (R_p) are given in table 2.

The corrosion potentials (E_{corr}), corrosion current density (i_{corr}), and anodic/cathodic Tafel slopes (b_a and b_c) were obtained from these results. Then, based on an approximate linear polarization at the corrosion potential, the polarization resistance(R_p) values were determined using the relationship [6, 11]:

$$R_p = \frac{b_a \times b_c}{2.303 \times i_{corr}(b_a + b_c)} \tag{1}$$

It can be seen that all PEO coatings and the black oxide coating had much lower i_{corr} compared to Zr-2.5Nb substrate (as much as one to two orders of magnitude) and the E_{corr} (-0.21V to -0.28V) are higher than the substrate (-0.35V). At the point E=0.78V, the i_{corr} of the Zr-2.5Nb substrate increased significantly. This is because the thin protective oxide film on the Zr-2.5Nb substrate surface broke down and large craters were formed (see insert in fig. 4).

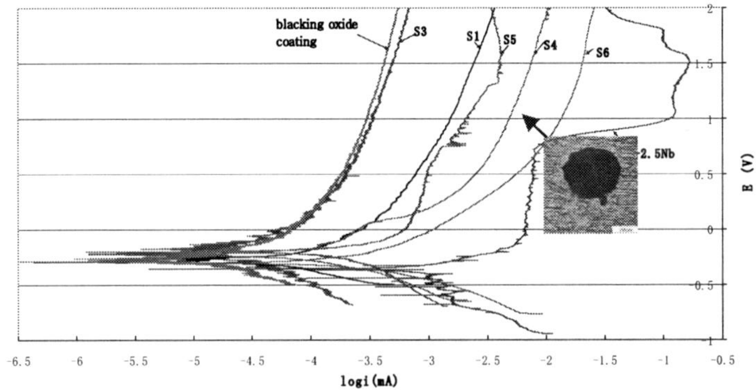

Figure 4: Potentiodynamic polarization curves of the uncoated Zr-2.5Nb, coated samples and black oxide coating.

Table 2: Results of potentiodynamic polarisation tests in a 0.2mol/L LiOH solution.

Samples	i_{corr}(mA/cm^2)	E_{corr}(V)	R_p(Ω cm^2)
Zr-2.5Nb substrate	1.02E-03	-0.35	1.22E+04
Black oxide coating	1.12E-05	-0.28	3.57E+06
S1 (0.05A/cm^2, 8:0.8)	6.31E-05	-0.25	1.22E+06
S2 (0.1A/cm^2, 8:0.8)	5.71E-05	-0.27	1.39E+06
S3 (0.25A/cm^2, 8:0.8)	1.58E-05	-0.22	1.58E+06
S4 (0.05A/cm^2, 8:8)	7.08E-05	-0.25	5.72E+05
S5 (0.1A/cm^2, 8:8)	1.58E-04	-0.21	2.68E+05
S6 (0.25A/cm^2, 8:8)	2.09E-04	-0.28	2.06E+05

From table 2, it can be seen that PEO coatings made in electrolyte I have a better corrosion resistance than those made in electrolyte II. It is in agreement with previous studies of the production of PEO coatings on Al, where alkaline electrolytes are widely used, but too high in PH value may have negative effect on the growth of passive films [5]. The thin commercial black oxide coating shows excellent corrosion resistance in these potentiodynamic polarization tests.

3.2.2 Autoclave experiment

Fig. 5 shows how the weight gain of samples changed with the exposure time in the autoclave. The weight gain of PEO coatings increased quickly in the first two days and then slowed down. This may explained by the dissolution of the oxygen from the solution into the outer layer of the PEO coating which is not very dense and forms ZrO_2 together with other intermediate lower oxides. However, there was a very thin, inner dense layer close to the oxide/metal interface which worked as an excellent barrier to the oxygen ingress, so the corrosion rate after 2 days slowed down. However, after exposure beyond a certain point (transition point), all coated samples had an abrupt increase in corrosion rate. Comparing two groups, it can be found that samples in Group A had a longer pre-transition period (in the range of 10 to 30 days) than those in Group B (in the range of 5 to 10 days). The weight gains of Group A were almost the same (0.76-0.87%). Due to the spalling of oxide layers in Group B samples, it was difficult to measure the weight gain after 10 days in autoclave. Thus, the weight gain data after the 10-day point in Group B shows a decrease compared to shorter exposure times.

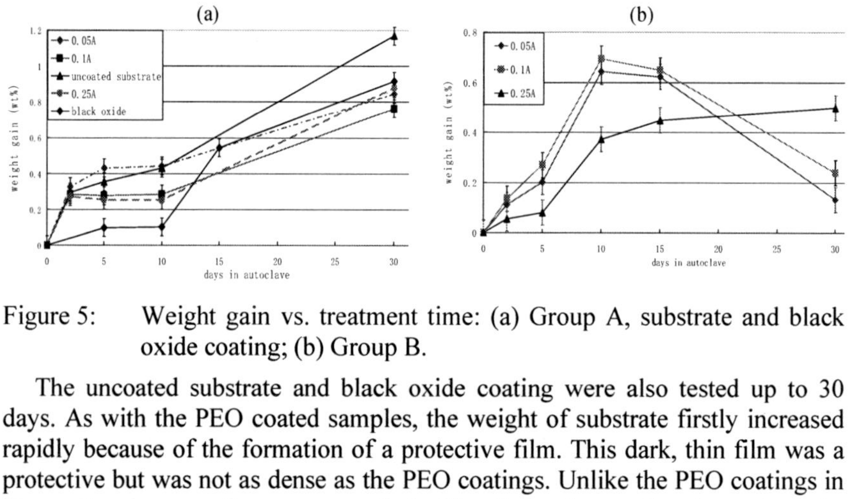

Figure 5: Weight gain vs. treatment time: (a) Group A, substrate and black oxide coating; (b) Group B.

The uncoated substrate and black oxide coating were also tested up to 30 days. As with the PEO coated samples, the weight of substrate firstly increased rapidly because of the formation of a protective film. This dark, thin film was a protective but was not as dense as the PEO coatings. Unlike the PEO coatings in Group B, where coatings were spalled off entirely, the coating formed on the uncoated substrate consists of several thin layers formed parallel to the oxide/metal interface [17]. When a certain thickness of oxide is attained, the outer layer of oxide will crack into small pieces, separate from the sample surface and sink to the bottom of the autoclave. In this case, the weight gain data recorded in Fig 5(a) for the substrate is much lower than its real value. However, the final weight gain (about 1.18%) is still higher than the PEO coatings. The black oxide coating shows an excellent corrosion prevention property in the first 10 days (better than all the PEO coatings). But, for long exposure times, it exhibited a similar weight gain to the PEO coatings (0.92%).

Fig. 6 is SEM micrographs showing surface morphologies of the two groups of PEO coatings and black oxide coating. Figs 6 (a) and (b) show the surface morphology of sample S1 after 5 and 30 days in the autoclave. The coating was still adherent and protective since its R_p value was still higher than the substrate (table 3). Unlike sample S1, sample S4 which was fabricated using a different electrolyte but the same current condition, behaved differently—almost all the coating flaked away from the metal (fig 6(c)). Instead of the PEO oxide, a newly formed oxide (fig 6 (c), (d)) quickly grew beneath the initial layer. This new oxide was white, soft and provided very little corrosion protection. After 30 day exposure, the PEO coating had spalled-off entirely, and a 'new' oxide, which was over 20μm thick, covered the whole surface (fig 6 (e), (f)). A thick, white oxide grew on the top of the black oxide coating (fig 6(g), (h) after 30 days in autoclave, but this oxide was different from the PEO coatings in its morphology. A flocculent structure with a large amount of pores was formed.

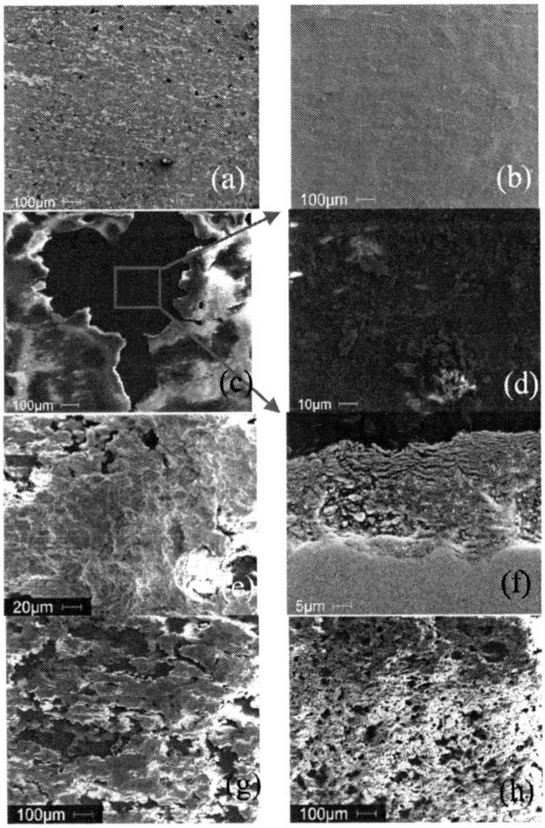

Figure 6: Morphology of the samples during the autoclave corrosion test under different magnifications: (a), (b) S1 after 5 and 30 days; (c)-(f) S4 in 15 and 30 days; (g), (h) inside surface of black oxide coating in 5 and 30 days respectively in autoclave.

Table 3 shows the potentiodynamic polarization results for specimens retested after the autoclave testing. From table 3, the thin oxide formed in first 5 days in 300°C aqueous environment acts as an excellent oxidation barrier but its corrosion resistance decreases with the exposure time after a critical time. It can be seen that with an increase of the exposure time, the value of i_{corr} increase greatly together with a slight increase of E_{corr}. In this case, the R_p for both coated and uncoated samples decreased by more than an order of magnitude. The R_p value of coatings of Group B decreased more rapidly than that in Group A. However, it is interesting to note that all PEO coatings had a better corrosion resistance than the 'newly-made' PEO ones after 5 days in autoclave. It was suggested that the small pores in 'newly made' PEO coatings were closed by heating in the autoclave for a short period of time.

Table 3: Results of the potentiodynamic polarization tests on samples removed from autoclave corrosion tests at fixed time intervals.

Days in autoclave	Samples	$i_{corr}(\mu A/cm^2)$	$E_{corr}(mV)$	$R_p(\Omega cm^2)$
0 day	Zr-2.5Nb	1.02	-330.2	1.22E+04
5 days	Zr-2.5Nb	0.02	-195.54	1.08E+06
	S1	0.015	-205.34	1.23E+06
	S2	0.028	-190.9	1.12E+06
	S3	0.019	-264.18	1.91E+06
	S4	0.029	-202.54	1.77E+06
	S5	0.029	-329.5	1.05E+06
	S6	0.025	-184.23	1.28E+06
10 days	S1	0.122	-251.9	3.38E+05
	S2	0.18	-229.1	2.61E+05
	S3	0.123	-245.4	4.4E+05
15 days	S4	0.206	-186.74	2.38E+05
	S5	0.3	-260.1	1.53E+05
	S6	0.128	-281.7	9.04E+04
30 days	Zr-2.5Nb	0.374	-265.3	9.21E+04
	S1	0.231	-251.9	1.22E+05
	S2	0.313	-127.2	9.31E+04
	S3	0.282	-171.6	1.0E+05
	S4	0.39	-221.8	5.3E+04
	S5	0.273	-193.6	4.98E+04
	S6	0.313	+30.451	6.0E+04

3.3 Pin-on-disc tests

The wear test results are summarized in table 4 for the rotating mode and table 5 for the reciprocating mode. Tables 4 and 5 show almost all the PEO coatings had not failed after a 1000m sliding distance in both rotating and reciprocating mode of testing. The black oxide coating, both under dry and lubricated conditions, showed inferior wear resistance. After wear testing, roughness data was collected across the wear trace to compare the mass loss (fig 7). A 50m wear distance for the Zr-2.5Nb substrate produced a 10μm-deep wear trace, whereas a 1000m wear distance under same load produced only a 1.5μm-deep trace for PEO-coated sample (S4). As mentioned in part **3.1**, the PEO coating (5μm thick) was not penetrated.

Table 4: Wear/failure distances of coated and uncoated samples under rotating mode.

Rotating mode	Zr-2.5Nb	S1	S2	S3	S4	S5	S6
dry/2N/ v=0.1m/s	50m	1000 m	970m (failed)	1000 m	1000 m	1000 m	910m (failed)

Table 5: Wear/failure distance of coated and uncoated samples under reciprocating mode.

Reciprocating mode	Zr-2.5Nb	Black oxide coating	PEO S1
dry/2N	×	40m (failed)	1000m
lubricated/5N/v=0.08m/s	50m	75m (failed)	1000m

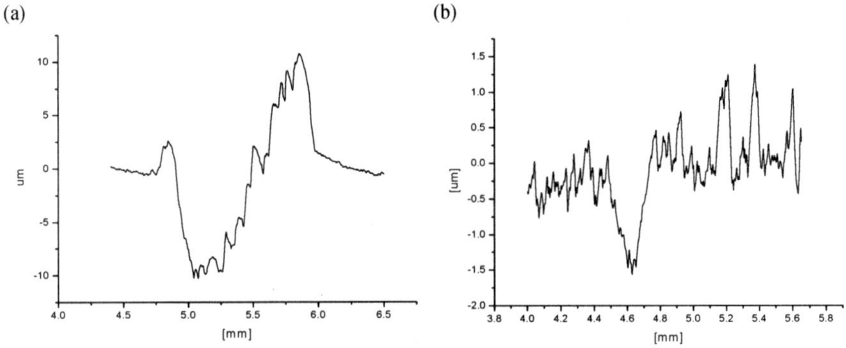

Figure 7: Transverse roughness on the wear trace of (a) Zr-2.5Nb substrate and (b) PEO coated sample S4 under 2N load, dry air, rotating mode with 50m and 1000m sliding distances respectively.

Fig 8 shows the wear traces for selected samples tested in the dry and rotating mode. 1000m wear distance was used on sample S1 after PEO process and after 5 and 10 days in autoclave (fig 8 (a)–(d)). Grinding polish was observed inside the wear trace of S1 (fig 8(b)) due to the high hardness of the PEO coating. The coating material had been compacted by the normal load. This behaviour was because the PEO layer had been hardened due to autoclaving for 5 days. However, after 10 days in the autoclave, the outer layer of PEO coating on sample S1 was softened (both by autoclave treatment and heat produced during wear tests) and was pushed away from the wear trace and became wear debris. However, the dense oxide underlayer was still intact. Fig 8 (c), (d) are the wear trace on sample S4 after 10-day autoclave treatment. The 1000m wear test caused the penetration of the PEO coating and the wear loss from the substrate. Inside the wear trace, deep scratches (grooves), caused by “three body” debris wear, were observed on the soft substrate material. The uncoated sample with 5-day autoclave treatment showed evidence of plastic deformations in its wear trace (fig 8 (g), (h)). Scaly debris was observed after 50m wear test which means the oxide formed in the autoclave was relatively soft.

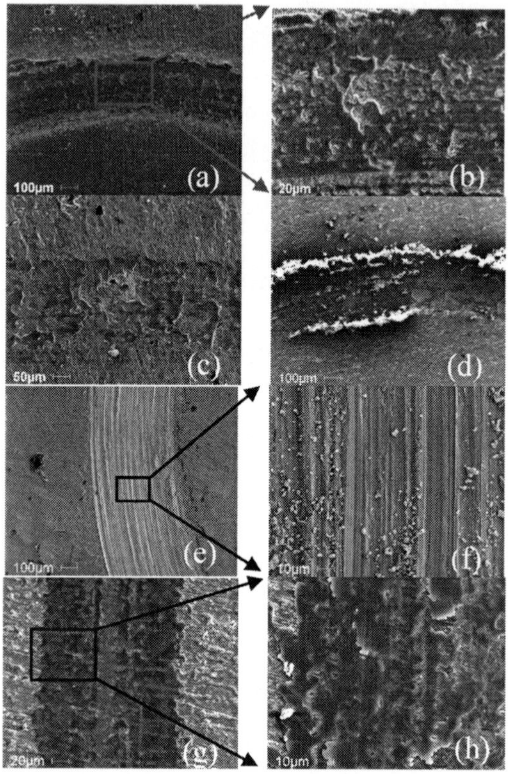

Figure 8: Rotating wear trace on different PEO coatings and substrate: (a, b) S1, 1000m, before autoclaved; (c), (d) S1 1000m, 5 and 10 days in autoclave; S4, 1000m, (e), (f)10 days in autoclave; (g), (h) Zr-2.5Nb substrate 50m, 5 days in autoclave.

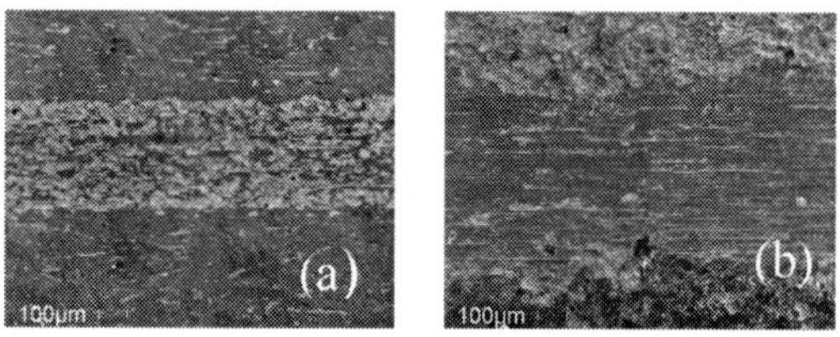

Figure 9: SEM showing wear trace of (a) sample S1, (b) black oxide coating.

After the rotating mode wear tests, sample S1 was chosen for comparison with the black oxide coating because of its excellent wear resistance. Because of the curved surface of black oxide coating sample, a reciprocating mode was used. The wear traces are shown in fig 9. 1000m and 75m wear distances were used on S1 and black oxide coating (failure distance), respectively, under the same 5N load and lubricated condition. The wear trace on black oxide coating was much wider and deeper than that on sample S1 and it is clearly seen that all coating material was removed from the substrate. Thus the PEO coating had much better wear resistance than the black oxide coating.

4 Conclusions

Investigations into the effects of process (electrolytic and electrical) parameters on PEO coating formation on Zr-2.5Nb alloy and the corrosion and wear properties of the coatings showed that:

1. For the limited PEO process parameters examined, the best coatings in terms of its integrity (dense, continuous) were produced at higher Na_2SiO_3 : KOH ratios (10:1 cf 1:1), i.e. lower PH values, and lower current densities, i.e. longer processing times.
2. The commercial, thin, black oxide coating showed the best corrosion resistance in the potentiodynamic polarization tests. PEO coatings also improved the corrosion resistance of the Zr-2.5Nb substrate (the polarization resistance (R_p) was one to two orders of magnitude higher than the substrate).
3. 5 μm thick PEO coatings have a much higher wear resistance than the commercial autoclaved black oxide coating under both dry and water lubricated conditions. Grinding polish was observed on the PEO coatings which enhanced the hardness of the coating surface. The commercially-produced black oxide coating had a wear resistance that was just slightly better than the substrate.
4. PEO coatings made from 10:1 of Na_2SiO_3: KOH proved to have much better corrosion resistance and lower weight gains than the Zr-2.5Nb substrate after 30 days in autoclave. The black oxide coating had a very low weight gain value in the first 10 days but at 30 days, the final weight gain was very close to the PEO samples made at 10: 1 electrolyte ratio. A white, soft oxide formed beneath the PEO coating, replacing it slowly, especially on PEO samples made at a 1:1 electrolyte ratio. The original thin black oxide coating turned to a thick, flocculent structure which had inferior in corrosion resistance as measured in potentiodynamic polarization tests after autoclave testing.

References

[1] Holt, R. A., In-reactor deformation of cold-worked Zr-2.5Nb pressure tubes. *Journal of Nuclear Materials.* **372** pp. 182-214, 2008.

[2] Warr, B. D., Elmoselhi, M. B., Newcomb, S. B., McIntyre, N. S., Brennenstuhl, A. M., & Lichtenberger, P. C., Oxide characteristics and their relationship to hydrogen uptake in Zirconim alloys, *Zirconium in the Nuclear Industry.* ASTM STP 1132, American Society for Testing and Materials, Philadelphia. pp. 740-757, 1991.

[3] Northwood, D., Meng-Burany, X. & Warr, B: Microstructure of Zr-2.5Nb alloy pressure tubing. *Zirconium in the Nuclear Industry.* ASTM STP 1132,

American Society for Testing and Materials, Philadelphia, American Society for Testing and Materials, Philadelphia, pp. 156-172, 1991.
[4] B. Cox; Some thoughts on the mechanisms of in-reactor corrosion of Zirconium alloys. *Journal of Nuclear Materials.* **171** pp. 331-368, 2005.
[5] Snizhko, L. O. & Yerokhin, A. L; Anodic processes in plasma electrolytic oxidation of aluminum in alkaline solutions. *Electrochimica Acta.* **49** pp. 2085-2095, 2004.
[6] Yerokhin A. L., Nie X., Leyland, A., Matthews A., Dowey S. J. Plasma electrolysis for surface engineering *Surface and Coating Technology.* **122** pp. 73-93, 1999.
[7] Barik, R. C., Wharton, J. A., Wood R. J. K., Stokes K. R. & Jones R. L. Corrosion, erosion and erosion-corrosion performance of PEO deposited Al_2O_3 coatings. *Surface & Coating Technology.* **199** pp. 158-167, 2005.
[8] Guang, Y. J., Xia, Y., & Li, G.; Growth mechanism and corrosion behaviour of ceramic coatings on Al produced by autocontrol AC pulse PEO. *Surface & Coatings Technology.* **202** pp. 4602-4612, 2008.
[9] Yerokhin A. L., Shatrov, A. Samsonov V., Shashkov P., Pikington A., Leyland A. & Matthew A. Oxide ceramic coatings on aluminium alloys produced by a pulsed bipolar plasma electrolytic oxidation process *Surface & Coatings Technology.* **199** pp. 150-157, 2005.
[10] Ghasemi, A., Raja V. S., Blawert C., Dietzel W. & Kainer K. U. Study of the structure and corrosion behaviour of PEO coatings on AM50 magnesium alloy by EIS *Surface & Coatings Technology.* **202** pp. 3513-3518, 2008.
[11] Yerokhin A. L, Nie, X., Leyland, A. &Matthews, A. Characterisation of oxide films produced by PEO of a Ti-6Al-4V alloy. *Surface and Coating Technology.* **130** pp. 195-206, 2000.
[12] Klapkiv, M. D. Povstyana N. Yu. & Nykyforchyn H. M. Production of Conversion Oxide-Ceramic Coatings on Zirconium and Titanium Alloys *Material Science*, **42(2)**, pp. 277-2862006.
[13] Pauporte T., Finne J., Kahn-Harari A., Lincot. D. Growth by plasma electrolysis of zirconium oxide films in the micrometer range. *Surface & Coating Technology*. **199**, pp:313-219, 2005.
[14] Bojinov, M., Karastoyanov, V., Kinnunen P. & Saario T. Influence of water chemistry on the corrosion mechanism of a zirconium–niobium alloy in simulated light water reactor coolant conditions. *Corrosion Science.* **52** pp. 54-67, 2010.
[15] Yilmazbayhan A., Breval E., Motta A. T. & Comstock R. J. Transmission electron microscopy examination of oxide layers formed on Zr alloys. *Journal of Nuclear Materials.* **349** pp. 265-281, 2006.
[16] Northwood D. O., & Kosasih U. Corrosion and Hydriding Behavior of Zr-2.5Nb Wt. Pct Nb Alloy Nuclear Reactor Pressure Tubing. *Journal Materials for Energy System.* **4(1)**, pp.3-15, 1982.
[17] Parfeno B. G., Gerasimov V. V. & Venediktova G. I., Corrosion of Zirconium and Zirconium alloys *IPST Press*, Wiener Bindery Ltd., Jerusalem. 1969.

Sliding mechanisms in tribological contact of TMD-C sputtered coatings

T. Polcar[1,2] & A. Cavaleiro[2]
[1]*Department of Control Engineering, Faculty of Electrical Engineering, Czech Technical University in Prague, Czech Republic*
[2]*SEG-CEMUC, Mechanical Engineering Department, University of Coimbra, Portugal*

Abstract

In this paper, results on the tribological behaviour of nanocomposite coatings composed by nanoplatelets of transition metal dichalcogenides (TMD) immersed in a C-rich amorphous matrix will be presented. It will be shown that this microstructure will allow low friction coefficients in wide range of surrounding environments, since TMD platelets are re-oriented in the C-matrix as a consequence of the mechanical energy liberated in the sliding contact. In this case, the oxidation of the TMD material is avoided due to the protection achieved by the C matrix. Examples of TMD-C coatings of the W-S-C and Mo-Se-C systems will be presented and discussed. After tribological characterization of the coatings in different environments under different loading conditions experimental evidence of the wear mechanisms will be given by cross section transmission electron microscopy, Raman spectroscopy and Auger electron spectroscopy.

Keywords: sliding mechanisms, self-lubrication coatings, low friction, TMD coatings.

1 Introduction

The idea behind the use of composite coatings, including different well-known phase materials with good tribological performance, was to achieve a synergetic effect that can overcome the individual drawbacks of each specific phase in low friction demands. One of the most interesting materials for the nanocomposite design of low friction coatings is any member of the transition metal

WIT Transactions on Engineering Sciences, Vol 66, © 2010 WIT Press
www.witpress.com, ISSN 1743-3533 (on-line)
doi:10.2495/TD100171

dichalcogenides (TMD) family. The weak Van der Waals forces between the lamellae of TMD allow an almost frictionless sliding, providing that it is performed in an absolutely clean environment, preferentially ultra-high vacuum [1]. Unfortunately, TMD is very vulnerable to environmental attacks; the presence of oxygen and water vapour hinders its excellent frictional properties.

Pulsed laser deposition and magnetron sputtering have been the most studied ways to prepare thin TMD films; the latter method is widely used in industrial scale for TMD coatings preparation [2]. Pure magnetron sputtered TMD films perform well in vacuum conditions, inert atmospheres or, to a less extent, in dry air. Nevertheless, even for such restricted application conditions, they suffer from a lack of adhesion to the steel substrates, low resistance to abrasive wear due to their low hardness and, generally, low load-bearing capacity. It should be pointed out that the optimum orientation of the TMD phase assuring a minimum friction, (002), is almost impossible to achieve except for very thin films [1, 3]. Therefore, a complicate process is required to achieve the low friction; in general, the TMD phase must be "reoriented", either by removal of the film material and consequent formation of a tribolayer or by reorientation of the TMD phases inside the carbon matrix.

On the other hand, the environmental attacks are difficultly overcome due to another typical feature of magnetron sputtered TMD films, their columnar structure, which allows the penetration of the oxygen or the water vapour inside the coatings [4, 5]. Finally, the adhesion and cohesion of such films is limited despite the improvement achieved with the use of metallic interlayers (mostly titanium).

To face the issues referred to above, several concepts were envisaged. One of the most successful solutions was alloying TMD with an additional metallic (Ti [2, 6], Pb [7], Au [8]) or non-metallic (C [9, 10], N [11]) elements. The main feature of the alloyed films was the improvement of the microstructure density, the enhancement of the hardness and, therefore, the increase in the load-bearing capacity.

In this paper, a new concept of TMD films co-sputtered with carbon based on the tailoring of the microstructure, which leads to excellent tribological properties, is presented.

2 New coatings design

In a first set of studies, W-S-C films were co-deposited by sputtering from a WS_2 target in a CH_4-containing reactive atmosphere. The flow of CH_4 and the target power were varied to prepare a large range of chemical compositions from pure WS_2 to W-S-C with carbon contents up to 70 at.%; the S/W ratio was about 1.2. Fig. 1(a) clearly shows that the main disadvantage of TMD films, the detrimental effect of the humid air, was not overcome and the friction in these conditions was relatively high whatever the carbon content was [9, 11]. However, the rapid film failure typical of pure tungsten disulphide was not observed and the coating wear rate was quite low (Fig. 1(b)), result attributed to the significant increase of the hardness, from 0.5 GPa (WS_2) to ~10GPa, achieved with the C incorporation.

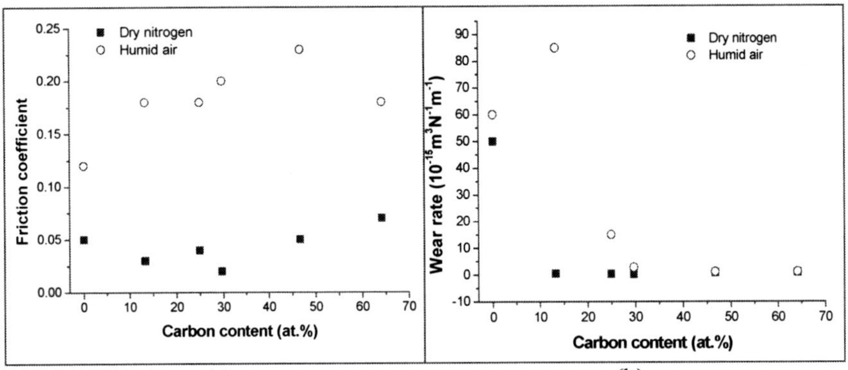

(a) (b)

Figure 1: Friction, (a), and wear, (b), coefficients of W-S-C reactive deposited sputtered coatings as a function of the C content when tested in humid air and dry N_2 (load 5N).

The analysis of the microstructure of the reactive deposited W-S-C coatings [5] allowed one to conclude that, for high carbon content, they were formed by a mixture of nanocrystals of WS_2 phase side by side with nanocrystals of W-C phases and small C-based amorphous zones (see example in Fig. 2), or just by WS_2 nanograins dispersed in the amorphous carbon matrix (low carbon content). A similar behaviour was observed elsewhere for WS_2/DLC/WC coatings showing a huge friction coefficient difference when sliding in humid or dry air [10]; the friction behaviour was attributed to the alternative formation of a WS_2-rich or a C-rich layer during the cyclic change of the environment from dry to humid, respectively. Another concept presented in literature for the TMD+C composite coatings deals with nanolayered structured coatings [12], where tungsten disulphide and carbon layers form a "super-lattice" film (the thickness of layers is in nanometre scale). The performance of such films was only moderate.

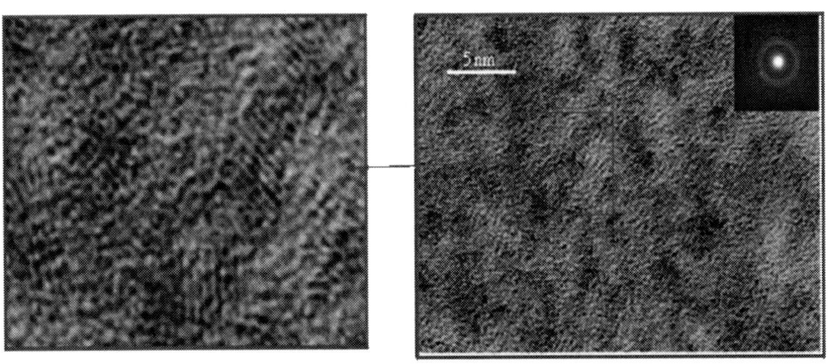

Figure 2: TEM micrographs of the reactive sputtered $W_{17}S_{15}C_{68}$.

WIT Transactions on Engineering Sciences, Vol 66, © 2010 WIT Press
www.witpress.com, ISSN 1743-3533 (on-line)

All these three different coating microstructures schematically shown in Fig. 3 proved to be unsatisfactory for reducing the friction coefficient in humid air sliding conditions. The multilayer films are vulnerable to cohesion damage, since interlayers sliding can occur in the bulk of the film where the shear stresses are larger than at the surface. This problem can be partially solved by thinning of layers down to units of nanometres; however, such a superlattice structure is difficult to be achieved on typical industrial substrates with standard surface roughness.

Figure 3: Schematic view of previous W-S-C films designs: Nanocomposite coatings with hard WC nanoparticles embedded together with WS_2 nanonograins in an amorphous carbon matrix (left); nanograins of WS_2 in a C-matrix (center); super-lattice WS_2+C layers film.

On the other hand, the other two microstructures cannot guarantee simultaneously that either the TMD nanocrystals could be re-oriented with the basal planes parallel to the sliding distance or the oxidation of the dangling bonds should not occur. In fact, during the re-orientation process the extremities of the grains will enter directly in contact with the atmosphere being immediately oxidized.

3 Approach

The basis for the achievement of a low friction coating in a great variety of environments could be envisaged by a micro(nano)structure where TMD nanocrystals could be deposited completely enrobed in a protective C-rich matrix. If possible, these nanocrystals should have shapes that allow an easy reorientation of the basal planes parallel to the sliding direction. C-matrix will contribute for improving the hardness and the compactness of the coatings, as well as for protecting the TMD against the oxidation until the moment that they are completely aligned and they arise to the contact surface. By their side, TMD should assure the low friction coefficients, due to the easy sliding of the basal planes, and only a few amounts of strong bonds, promoted by the oxygen, should exist. In parallel with this approach, the use of other types of TMDs with less sensibility to the oxidation, such as $MoSe_2$ [13], should also be envisaged.

During sliding in pin-on-disk testing, strong shear stresses occur which could contribute for the re-orientation of the TMD crystals in the C-based amorphous matrix. However, during the alignment of the TMD nanocrystals parallel to the sliding direction, their contact with the atmosphere gives rise to the reaction of the dangling bonds of the end plans with oxygen with the consequent increase of the friction. In pure TMD coatings this phenomenon was not observed due to the porous structure and very low hardness giving rise to very high wear rate [3]. Therefore, as generally accepted in the wear models it is expected that TMD material is worn out and then adheres on both surfaces in the contact. Such tribolayer is very thin (nanometre scale) with highly oriented TMD phase, mainly with the basal plans parallel to film surface [1, 3].

The problems referred to above could be avoided if the crystals were already aligned in the material, since in this case almost no dangling bonds would be available for reaction [14]. Recently, it was possible to prove that nanocrystals of a compound, such as TiC, could be re-arranged through the C-matrix where they were embedded, during the mechanical deformation of the material by indentation [15]. It could be possible that TMD-C films, where nanograins of TMD are embedded in an amorphous carbon matrix, exhibit similar behaviour. Nevertheless, the re-orientation of TMD crystals in a carbon matrix has not been studied yet.

Thus, it was decided to prepare TMD-C films with very small nanograins of the TMD phase (reduced, if possible, to one cell of TMD), since small grain size could promote more easily the re-orientation of the TMD phase inside the carbon matrix.

4 How to deposit the coatings with this nanocomposite microstructure?

In order to deposit coatings containing simultaneously C and the TMD phase with a new microstructure, the deposition by co-sputtering from TMD and C sources, either from separate targets or by a puzzled target with both materials (Fig. 4), can be an alternative.

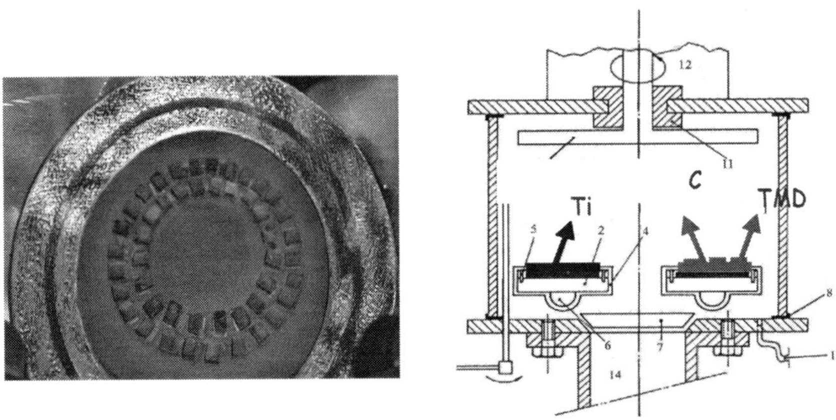

Figure 4: Schema of the deposition chamber with two electrodes and picture of the puzzled target. For separate target deposition, the Ti target is replaced by a TMD one whereas in the other electrode only a C target is attached.

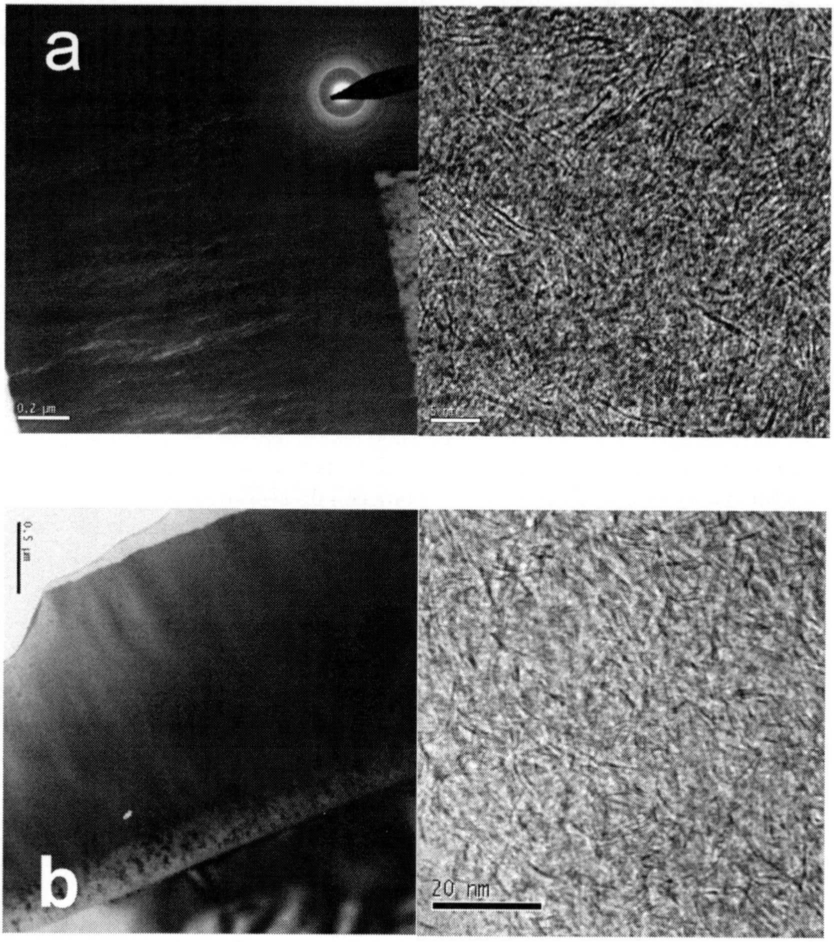

Figure 5: TEM and HR-TEM images of W-S-C (a) and Mo-Se-C (b) co-sputtered films deposited with ~50 at.% C. Note the dense structure (left) and the separated TMD platelets (right).

With this geometry, microstructures such as those shown in Fig. 5 for Mo-Se-C and W-S-C films with ~50 at.%C can be deposited [16, 17]. The TMD phase is under a platelet shape with only a couple of stacked basal planes. The platelets are completely enrobed by the C-matrix. X-ray diffraction and Raman spectroscopy undoubtedly confirmed the presence of the TMD and the C phases [18–20]. The hardness was about 3 and 10 GPa for Mo-Se-C and W-S-C, respectively, compared to the <0.5 GPa of the pure TMD films sputtered under identical conditions from the TMD target.

5 Friction performance

Figures 6 a) and b) show the friction coefficient obtained in a sliding reciprocating pin-on-disk test (frequency - 20 Hz; stroke length - 1 mm) for two coatings of the systems W-S-C and Mo-Se-C with approximately 50 at.% C. Two facts are evident: (i) in humid conditions Mo-Se-C coatings show better friction performance than W-S-C ones; (ii) in some loading conditions (very high applied loads) the friction coefficient in humid air approaches that in dry conditions. This last result gives strong evidence that the scope of re-orientation of the TMD nanocrystals, in conditions of oxidation protection, can be achieved. However, important amounts of mechanical energy have to be used to allow the correct reorientation in the envisaged conditions. The progressive decrease of the friction coefficient with increasing applied load suggests that the sliding process is driven by the TMD phases in the contact (such decrease is typical of TMD [21]), since pure carbon films identical to the amorphous carbon matrix of these coatings exhibit a slight increase of the friction with load [16]. It is also important to remark that a similar behavior is found in dry conditions although with a starting value much lower (~0.08 against ~0.20 for a 20 N load).

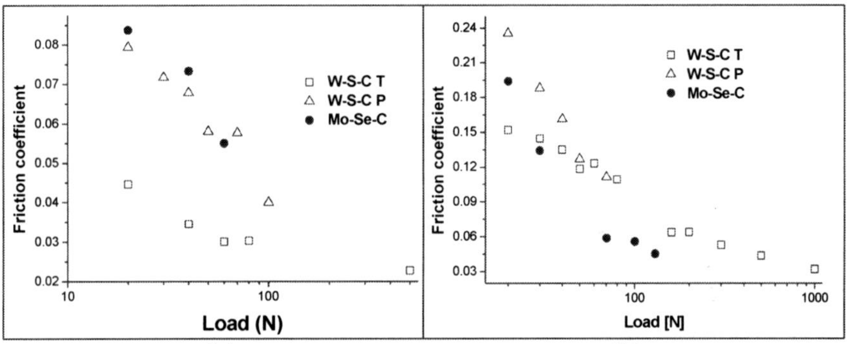

(a) (b)

Figure 6: Friction coefficient of TMD-C coatings as a function of the applied load in (a) dry and (b) humid conditions. Note logarithmic load scale.

Figure 7 shows the average friction coefficient obtained in tests with increasing duration times when a very low load (5 N) is applied to the ball. As can be confirmed, there is a steep decrease of the friction coefficient in the first thousands of cycles being possible to reach a steady state value (~0.05) close to 100,000 cycles, which is not far from the lowest value showed in figure 6 for these coatings when tested with much higher loads. In another paper of the authors, it was shown the evolution of the friction coefficient as a function of the number of cycles for a W-S-C coating tested by changing the humidity in the atmosphere every 10,000 cycles [22]. Besides the progressive approach between the friction values in both, dry and humid, environments with increasing test duration, also in both cases a decreasing trend in each step could be observed.

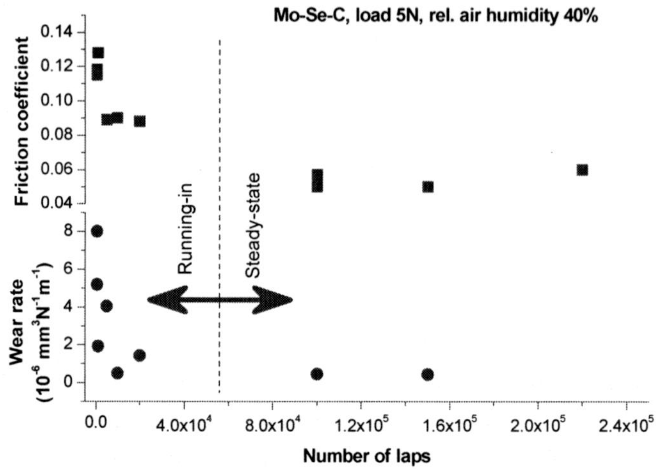

Figure 7: Evolution of the friction coefficient and the wear rate after tests with different duration of a Mo-Se-C coating sliding under room conditions with an applied load of 5 N (pin-on-disc, 100Cr6 ball with a diameter of 6 mm). The steady state wear regime was reached after long running-in.

In conclusion, in sliding tests of TMD-C coatings, low friction coefficients can be achieved in any environment provided that sufficient contact pressure or test duration are used.

6 Evidence of the sliding mechanism

The progressive way as to how the formation of the TMD tribolayer occurs can be demonstrated experimentally by the analysis of the top surface of the sliding track after wear testing. Figure 8 shows the Raman spectra of TMD-C films obtained after tribological tests carried out with either the same number of cycles but different loads or the same load but different number of cycles. It is clear that the increase in the load and/or the number of cycles gives rise to a more intense signal of the bands assigned to the TMD phase in comparison to those of the C-based material (G and D bands). This result confirms an increasing agglomeration of TMD material close to the contact zone.

In previous studies [16, 17], Auger spectroscopy (AES) showed that the TMD signal increase in Mo-Se-C coatings was due to the thickening of the surface TMD tribolayer when contact pressure and/or number of laps (i.e. test duration) was increased. It was clear the absence of the C-signal on the top layer suggesting that this was formed almost exclusively by TMD. However, the agglomeration of the TMD material in the contact zone should only lead to the reduction of the friction coefficient providing that its basal planes are aligned parallel to the surface and no abundant oxidation occurs that can lead to strong bonding between them. It is therefore also important to remark the absence of O in this top layer (compare to the residual signal of the material underneath)

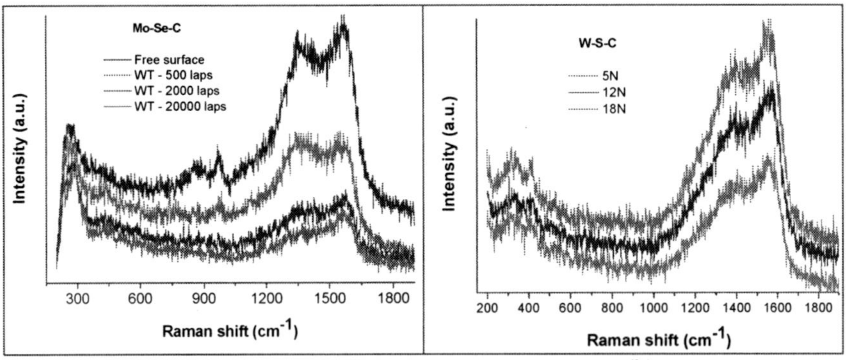

(a) (b)

Figure 8: Raman spectra taken from the center of the wear track as a function of the test duration (a) and applied load (b).

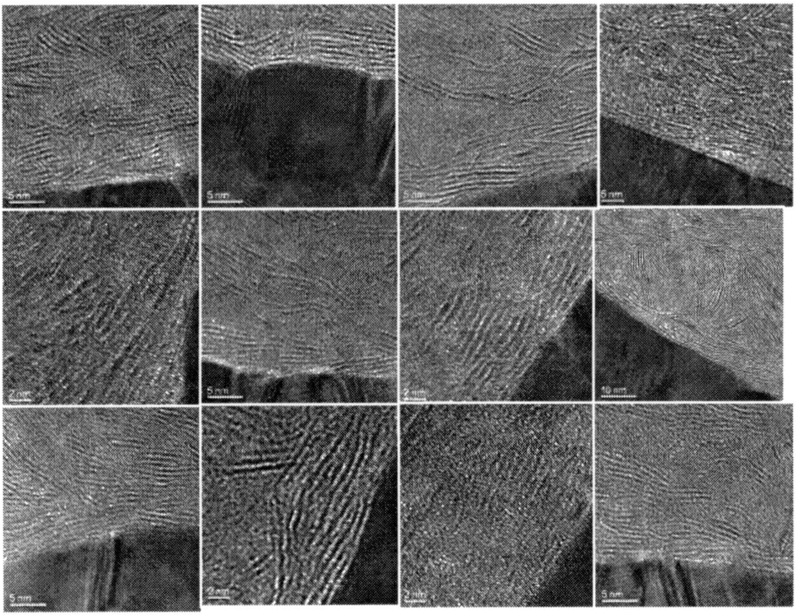

Figure 9: HR-TEM images of the cross-section of the wear tracks prepared by Focused Ion Beam. The black in the images is the protective gold layer sputtered over the wear track. The preferential orientation of the TMD platelets is clearly visible contrasting to as-deposited state shown in Fig. 5.

stating that the formation of the tribolayer was achieved in a protective way, without the oxidation of the dangling bonds of the original short TMD platelets.

The microstructural observation of the cross section of the top layer of the wear track by HRTEM (high-resolution transmission electron microscopy) showed that the TMD material was aligned parallel to the surface covering the most part of the zones which were analyzed, see Fig. 9. Furthermore, the same micrograph shows that as the analysis is being moved to the interior of the film, the alignment is less and less evident. The observation of a great number of micrographs of different zones of the wear track [17] permitted concluding that the reorientation of the TMD platelets is made progressively from some tens of nanometers until the track surface. Almost no other orientations but parallel to the surface are encountered in the first 5 nm top layer.

7 Conclusions

The deposition of nanocomposite coatings of the W-S-C and Mo-Se-C systems by co-sputtering from $WS_2/MoSe_2$ and C materials originated a nanocomposite structure composed by $WS_2/MoSe_2$ platelets embedded in a C-rich amorphous matrix. With this morphology, during the tribological testing the nano-platelets could be re-oriented parallel to the sliding direction giving rise to low friction coefficients. It was demonstrated that the C-matrix had two roles: (i) it contributed to the increase of the hardness and load bearing capacity and (ii) it avoided the oxidation of the TMD platelets during re-orientation permitting low friction coefficients in O-containing environments. In order to achieve a correct re-orientation of the TMD platelets, enough mechanical energy was required, explaining why the friction coefficient progressively decreased when the duration of the test increased, at the same applied load, or the load increased (for constant test durations). The experimental evidence of the friction performance was given by high resolution transmission electron microscopy, Raman spectroscopy and Auger spectrometry.

Acknowledgements

This work was supported by the Czech Science Foundation through the project 106/07/P014, by the Ministry of Education of the Czech Republic (project MSM 6840770038), by EU through the FOREMOST project and by the Fundação para a Ciência e a Tecnologia through the project SFRH/BPD/34515/2006.

References

[1] Lansdown, A.R., *Molybdenum Disulphide Lubrication*, Elsevier: Amsterdam, 1999.

[2] Renevier, N.M., Fox, V.C., Teer, D.G., Hampshire, J., Coating characteristics and tribological properties of sputter-deposited MoS2/metal composite coatings deposited by closed field unbalanced magnetron sputter ion plating *Surface Coatings Technology*, **127(1)**, pp. 24-37, 2000.

[3] Moser, J., Lèvy, F., MoS2-x lubricating films - structure and wear mechanisms investigated by cross-sectional transmission electron-microscopy. *Thin Solid Films,* **228(1-2)**, pp. 257 -260, 1993.

[4] Regula, M., Ballif, C., Moser, J.H., Lévy, F., Structural, chemical, and electrical characterisation of reactively sputtered WSx thin films. *Thin Solid Films,* **280(1-2)**, pp. 67-75, 1996.

[5] Nossa, A., Cavaleiro, A., Chemical and physical characterization of C(N)-doped W-S sputtered films. *Journal of Materials Research*, **19(8)**, pp. 2356 -2365, 2004.

[6] Teer, D.G., New solid lubricant coatings. *Wear*, **251**, pp. 1068-1074, 2001.

[7] Wahl, K.J., Seitzman, L.E., Bolster, R.N., Singer, I.L., Low-friction, high-endurance, ion-beam-deposited Pb-Mo-S coatings. *Surface Coatings Technology*, **73(3)**, 152-159, 1995.

[8] Simmonds, M.C., Savan, A., Pfluger, E., Van Swygenhoven, H., Mechanical and tribological performance of MoS2 co-sputtered composites. *Surface Coatings Technology* **126(1)**, pp. 15-24, 2000.

[9] Evaristo, M., Nossa, A., Cavaleiro, A., W-S-C sputtered films: Influence of the carbon alloying method on the mechanical properties *Surface Coatings Technology*, **200(1-4)**, pp. 1076-1079, 2005.

[10] Voevodin, A.A., O'Neill, J.P., Zabinski, J.S., Nanocomposite tribological coatings for aerospace applications. *Surface Coatings Technology*, **116**, pp. 36-45, 1999.

[11] Nossa, A., Cavaleiro, A., Tribological behaviour of N(C)-Alloyed W-S films. *Tribology Letters*, **28(1)**, pp. 59-70, 2007.

[12] Noshiro, J., Watanabe, S., Sakurai, T., Miyake, S., Friction properties of co-sputtered sulfide/DLC solid lubricating films. *Surface Coatings Technology*, **200(20-21)**, pp. 5849-5854, 2006.

[13] Kubart, T., Polcar, T., Kopecký, L., Novák, R., Nováková, D., Temperature dependence of tribological properties of MoS2 and MoSe2 coatings. *Surface Coatings Technology*, **193(1-3)**, pp. 230-233, 2005.

[14] Evaristo, M., Polcar, T., Cavaleiro, A., Tribological behaviour of C-alloyed transition metal dichalcogenides (TMD) coatings in different environments. *International Journal of Mechanics and Materials in Design*, **4**, pp. 137-143, 2008.

[15] Chen, C.Q., Pei, Y.T., Shaka, K.P., De Hosson, J.M.T., Nanoscale deformation in TiC/a-C multilayered nanocomposite coatings. *Applied Physics Letters*, **92(24)**, 241913, 2008.

[16] T. Polcar, M. Evaristo and A. Cavaleiro, Can W-Se-C coatings be competitive to W-S-C ones?. *Plasma Processes & Polymers*, **6**,DOI: 10.1002/ppap.200930414

[17] Polcar, T., Evaristo, M., Colaço, R., Sandu, C. Silviu, Cavaleiro, A., Nanoscale triboactivity: The response of Mo-Se-C coatings to sliding. *Acta Materialia*, **56(18)**, pp. 5101-5111, 2008.

[18] Polcar, T., Evaristo, M., Cavaleiro, A., The tribological behavior of W-S-C films in pin-on-disk testing at elevated temperature. *Vacuum*, **81(11-12)**, pp. 1439-1442, 2007.

[19] Polcar, T., Evaristo, M., Cavaleiro, A., Friction of self-lubricating W-S-C sputtered coatings sliding under increasing load. Plasma Processes & Polymers, **4**, pp. S541-S546, 2007.
[20] Polcar, T., Evaristo, M., Stueber, M., Cavaleiro, A., Synthesis and structural properties of Mo-Se-C sputtered coatings. *Surface Coatings Technology*, **202(11)**, pp. 2418-2442, 2008.
[21] Singer, I.L., Bolster R.N., Wegand, J., Fayeulle, S., Stupp, B.C., Hertzian stress contribution to low friction behavior of thin MoS_2 coatings. *Applied Physics Letters*, **57(10)**, pp. 995-997, 1990.
[22] Polcar, T., Evaristo, M., Cavaleiro, A., Comparative study of the tribological behavior of self-lubricating W-S-C and Mo-Se-C sputtered coatings. *Wear*, **266(3-4)**, pp. 388-392, 2009.

Section 4
Tribology under extreme conditions

Wear and friction modelling on lifeboat launch slipway panels

B. Thomas[1], M. Hadfield[1] & S. Austen[2]
[1]*Sustainable Design Research Centre, School of Design, Engineering and Computing, Bournemouth University, UK*
[2]*Royal National Lifeboat Institution, Poole, UK*

Abstract

The Royal National Lifeboat Institution provides search and rescue cover along the United Kingdom and Republic of Ireland coast using a variety of lifeboats and launch techniques. In locations where there is no natural harbour it is necessary to use an inclined slipway to launch the lifeboat into the sea. Slipway stations consist of an initial section where the boat is held on rollers followed by an inclined keelway lined with low friction composite materials, the lifeboat is released from the top of the slipway and proceeds under its own weight into the water. The lifeboat is subsequently recovered to the top of the slipway using a winch line. With the introduction of the new, larger 'Tamar' class lifeboat it is necessary to upgrade existing boathouses and standardise slipway operational procedures to ensure consistent operation. The higher contact pressures and launch velocities associated with the new lifeboat have led to issues of high friction and wear on the low friction composite linings and a number of methods have been adopted to mitigate this effect. This paper presents a methodology for assessing slipway lining performance so that friction and wear conditions along the slipway can be monitored to ensure consistent operation. A multidisciplinary approach using tribometer testing in conjunction with finite element analysis and real world slipway condition surveys is adopted to extend the scope of investigation to incorporate common real world effects such as panel misalignments. Various lubricants are assessed for their suitability with regard to friction and wear performance in addition to sustainability considerations using the methodology, and modifications to the design of slipway panels, guidelines for lifeboat operation procedures and suitable panel installation tolerances are developed. Finally, new slipway condition monitoring procedures are proposed

WIT Transactions on Engineering Sciences, Vol 66, © 2010 WIT Press
www.witpress.com, ISSN 1743-3533 (on-line)
doi:10.2495/TD100181

incorporating slipway panel failure and replacement criteria and recovery winch based condition monitoring.

Keywords: lifeboats, tribology, wear, friction, slipway.

1 Introduction

The RNLI provides marine search and rescue cover to the UK and Ireland; this is achieved using a range of lifeboats and lifeboat stations positioned around the UK according to local conditions. If possible a natural harbour is used to store the lifeboat when not in use, however, at certain points around the UK there is no suitable natural harbour located to allow continuous coastal cover within the range of current lifeboats; here a slipway station is used. Lifeboat slipway stations use an inclined slipway with the lifeboat kept in a large boathouse at the top of the slipway, to launch, the lifeboat is released from the boathouse and proceeds under its own weight down the slipway into the water.

The RNLI operates 18 slipway stations in the UK and 2 in the Republic of Ireland, each station is different and each has its own historical launch techniques and slipway geometries though they all follow a common template [1]. The most common configuration is to have an upper section of steel rollers leading onto a smooth, lined, lower section. The boat is launched down the slipway with the plane slipway section coated with grease and recovered by manoeuvring the lifeboat onto the bottom of the slipway before attaching a winch cable and hauling the boat to the top of the slipway. Slipway inclines rage from 1 in 5 to 1 in 6 with most using a gradient of 1 in 5.The lower section of the slipway has been lined with a number of different materials through the years and currently utilises a graphite infused jute/phenolic resin composite selected for its low friction coefficient.

A variation on this slipway design uses steel rollers along the whole length of the slipway and a shallower slipway incline for areas with shallow beaches. This configuration works well for launches but can be very difficult to operate during the lifeboat recovery phase due to the difficulty in locating the lifeboat on the steel rollers prior to recovery. One innovative solution to this problem is to use

Figure 1: Slipway station variations – left, typical configuration at Padstow, right, separate launch and recovery slipways at Sennen Cove.

separate slipways for launch and recovery as at Sennen Cove (fig. 1) with a rotating cradle to transfer the lifeboat back onto the launch slipway once in the boathouse.

Improvements in lifeboat safety, instrumentation, speed and range have led to a steady increase in lifeboat size and mass and this has meant significant upgrading to slipways and boathouses through the years [2]. The Tamar is the latest slipway lifeboat design to be adopted, being introduced from 2006 and is the largest, heaviest slipway lifeboat yet used [3].

This has presented serious problems for slipway station operation with slipway contact pressures and winch loads increasing significantly. This has manifested as problems during lifeboat recovery with increasing winch loads, high friction and high wear along the plane, lined slipway section. Historically problems of high friction during launch or recovery have been addressed by manually applying marine grease to the slipway, and this has become common practice when operating the predecessor to the Tamar, the Tyne. The widespread manual application of marine grease on lifeboat slipways has implications for crew safety, cost and for the local environment and alternative solutions to high friction issues are to be sought. Similarly, higher contact pressures at the contact of the Tamar keel on the slipway lining in comparison to it predecessor have led to severe wear observed on the composite slipway lining panels. This has meant that composite panels designed to last for around two years are requiring replacement in a matter of months or even weeks. The composite panels are expensive, and at current replacement rates the panels are estimated to cost the RNLI around £260,000 per year [4] once the Tamar has been fully deployed to all slipway stations.

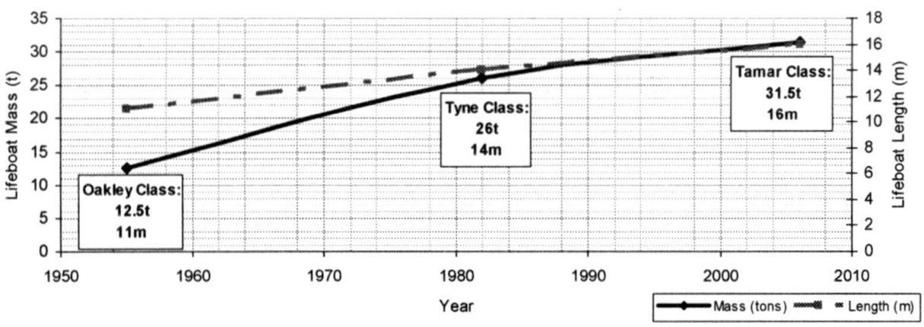

Figure 2: Slipway lifeboats mass and length over time.

2 Methodology

2.1 Initial investigations

The lifeboat slipway problem is investigated using a multidisciplinary approach. Initial investigations focus on site surveys and detailed investigation of worn slipway panels with additional friction data provided by RNLI line pull trials [5]. These showed characteristic wear patterns on the slipway panels and also indicated that there was significant misalignment of up to 4mm in height between panels. Initial concerns were raised as to the effects of these panel misalignments on the lifeboat keel contact.

Slipway line pull trials involve placing a load cell in line with the recovery winch to measure the friction along the slipway as the boat is hauled into the boathouse. These are regularly conducted by the RNLI at all slipway station to monitor slipway and winch performance. The data from these trials was investigated for any indications regarding the performance of the slipway lining panels.

2.1.1 Slipway geometry analysis

In addition to the site surveys and line pull data collected it was necessary to investigate the slipway geometry to derive an appropriate friction specification. It is found that for the most common configuration using a slipway incline of 1 in 5 the maximum friction coefficient is 0.2. For the shallower 1 in 6 configuration this falls to 0.167. However, the winch specification in all cases is for a 12 tonne maximum winch load, in the 1 in 5 gradient slipway case this reduces the maximum friction coefficient specification to 0.15. For the 1 in 6 gradient slipways the maximum friction coefficient required by the winch is $\mu=0.181$. This prompts the immediate recommendation that the recovery winch be specified according to the slipway incline, increasing the winch specification for the 1 in 5 incline from 12 tonnes to 13.7 tonnes allows the maximum friction specification for the slipway lining material to rise from 0.15 to 0.2. Conversely, the winch specifications for the 1 in 6 gradient slipway stations can be relaxed to 11.5 tonnes to match the geometric friction coefficient of $\mu=0.167$.

The current composite slipway lining has a listed dry sliding friction coefficient of $\mu=0.25$ which implies that it must be used with a lubricant in all cases.

2.2 Initial experimental testing

To gain a more complete picture of the contact mechanisms at work in this case samples of the slipway lining and the steel keel are tested using rotary and reciprocating tribometers. The friction and wear performance of the contact pair under various contact pressures and lubrication regimes is tested with initial screening friction and wear tests conducted on a modified Plint TE57 [6] under a pin on plate configuration with a lubricant bath to contain the lubricant. The Plint TE92 rotary tribometer [7] is used in a modified ring on ring arrangement to perform extended wear tests under varying contact pressure and lubricants.

Tests were conducted at contact pressures from 60-740kPa with the real world contact pressure calculated as 148kPa, the upper contact pressure limit tested represents five times this value. Tests were conducted under dry sliding conditions, using the marine grease currently used on most slipways, using freshwater and seawater lubrication, and using alternate lubricants – two biodegradable greases and a microsphere emulsion, originally intended as a cable pulling lubricant and used on the Sennen Cove recovery slipway.

The results from the testing indicate a number of points, the first being that the wear generated by the full lifespan rotary tests, designed to replicate the sliding experienced during the two year design life of the composite panels, was very low under all lubrication conditions and insufficient to explain the high panel failure rates experienced at Tenby and Padstow. This indicates that real world wear is likely to occur under different conditions than those tested here.

The second point raised by the testing confirms that the friction performance under dry sliding conditions is insufficient to meet the criteria identified above. This prompts the recommendation to discontinue any unlubricated running of lifeboat slipways. All lubricants tested performed sufficiently well to allow safe running of the lifeboat, although one of the biogreases tested and both water lubricated cases are close to the friction coefficient limit.

The presence of graphite in the composite indicates that a transfer layer is likely to form under dry sliding conditions, and this is seem in the smooth, reflective surface shown in fig. 3, this is also shown when the surface is inspected using profile inferometry with the roughness of the dry sliding wear scar markedly less than the virgin material. The presence of lubricants in the

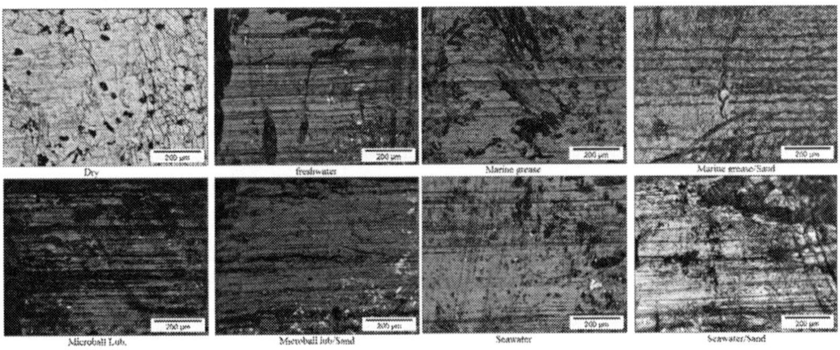

Figure 3: TE57 reciprocating tribometer: wear scar images, L-R dry, freshwater, marine grease, marine grease/sand, microball lubricant, microball lubricant/sand, seawater, seawater/sand.

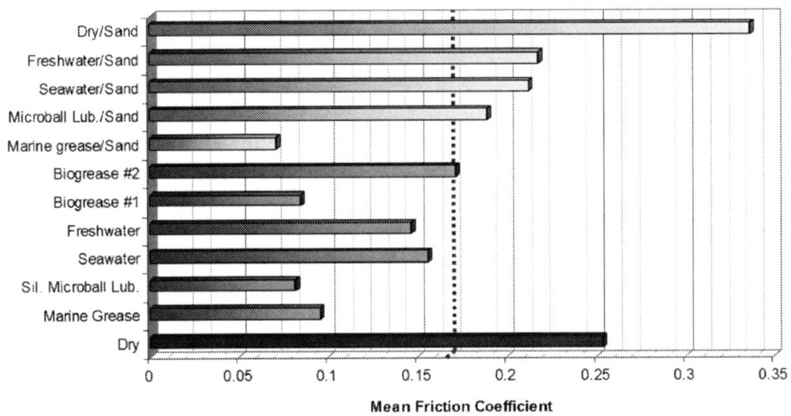

Figure 4: TE57 reciprocating tribometer: friction results.

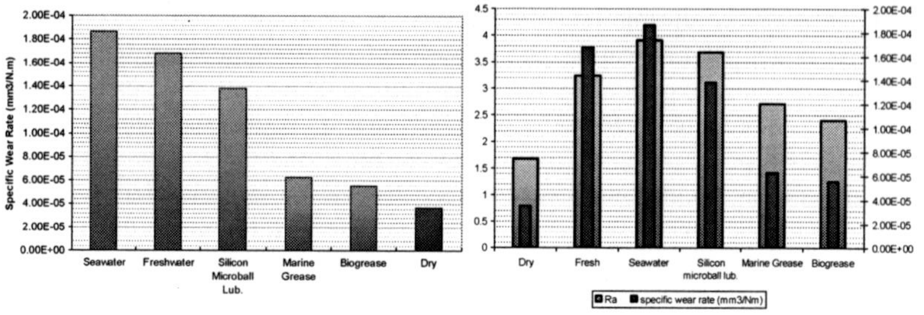

Figure 5: TE92 rotary tribometer: wear and surface roughness results.

contact region is known to interfere with the formation of a graphite contact layer [8] and this is seen in the partially smooth surfaces shown under lubricated conditions.

We would expect the wear rate of a sliding contact to be reduced by the presence of lubricants, however in this case the dry sliding wear rate is mitigated by the presence of a transfer layer and is particularly low, while the presence of lubricants acts to disrupt this transfer layer leading to the increase in wear rates in the presence of lubricants. To check this, the surface roughness of the wear scar is recorded and compared to the dry sliding case as an indication to the extent of the graphite transfer layer development. In all cases the surface roughness is higher in the presence of a lubricant which indicates that the smoothing effect of a graphite transfer layer is less developed, the surface roughness and wear rates show a good correlation indicating that the transfer layer performs an important role in reducing dry sliding wear. It should be noted however that the wear rate recorded in all cases is low, and not sufficient to explain the high panel failure rates encountered on real world slipways.

2.3 Further analysis

Following the tribometer wear results which indicated that the wear rates under aligned conditions are insufficient to explain the failure rates seen in real life, the scope of the research was developed using Finite Element analysis in ANSYS [5]. This was used initially to model the aligned case, where wear scar development was seen to correspond well with the real world case, and also to investigate the likely effects of the panel misalignments seen in the slipway surveys. These misalignments can be split into three main types, parallel offset, where one panel is raised in reference to its predecessor while remaining parallel, angled offset, where one end of the panel is raised in relation to its neighbour and the raised panel is pitched at an angle along its length, and skewed misalignment, where one panel is angled longitudinally in regard to its predecessor, corkscrew fashion. These scenarios are each tested for panel offsets of 0-5mm as indicated by the slipway surveys conducted.

The results show that the effects of panel misalignment are significant, as the contact moves away from the evenly distributed case the contact stresses and consequently wear, increase dramatically. The increase in contact stresses at the

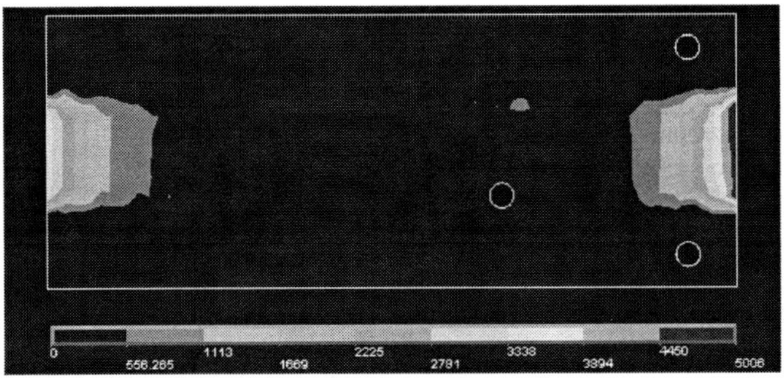

Figure 6: FEA: Friction map for misaligned panel (raised edge on the right).

raised regions is also a major factor in the friction performance of the panel, with panel friction maps also showing increases in frictional forces at the raised panel regions.

Table 1: Panel misalignment apparent friction coefficient contribution.

Offset	Parallel	Angled	Skewed
0	0.028	0.028	0.028
1	0.503	0.247	0.164
2	1.035	0.492	0.368
3	1.565	0.742	0.582
4	1.752	0.984	0.827
5	2.194	1.215	1.042

The end result of this is that the slipway recovery winch encounters higher friction forces than would be expected for an ideally aligned plane case, this can be interpreted as an increase in the *apparent* friction coefficient (although in reality the friction coefficient remains the same and the increase in frictional forces is a consequence of misrepresenting the contact case).

Wear is noted during real world surveys to occur most severely on misaligned slipway panels and this is reflected in the FEA models, with good correlation between real and simulated wear scars found.

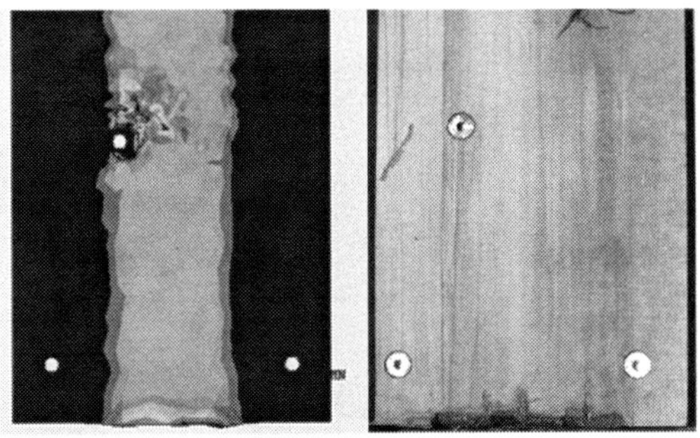

Figure 7: FEA: Simulated vs. real world wear on a misaligned panel.

These results indicate that it is likely that panel misalignments play an important role in both the friction and wear of real world slipway panels and this allows development and testing of a panel design that will reduce the effects of panel misalignments.

2.3.1 Solution proposed

The panel misalignments found during the slipway condition surveys were up to a 5mm step, however, it is seen from this analysis that even small panel misalignments can still have a serious effect on friction and wear; in fact the largest offset height that can be accommodated within the plastic deformation of the composite panel is just 0.361mm. Because of this, and because of the practical difficulties in reducing real world panel misalignments below this level it is decided to focus on techniques to limit the effect of panel misalignments, rather than attempt to eliminate the misalignments themselves. A panel modification to feature a chamfer at each end is proposed to reduce the stress concentration effects at the raised edges of panel misalignments and their consequent effects on panel friction and wear performance.

The aim of this panel is to allow reliable friction and wear performance along the slipway within current panel alignment standards.

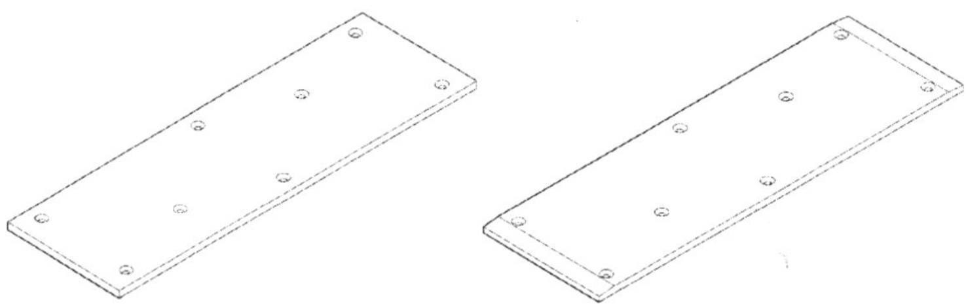

Figure 8: Original (left) and modified (right) slipway panels.

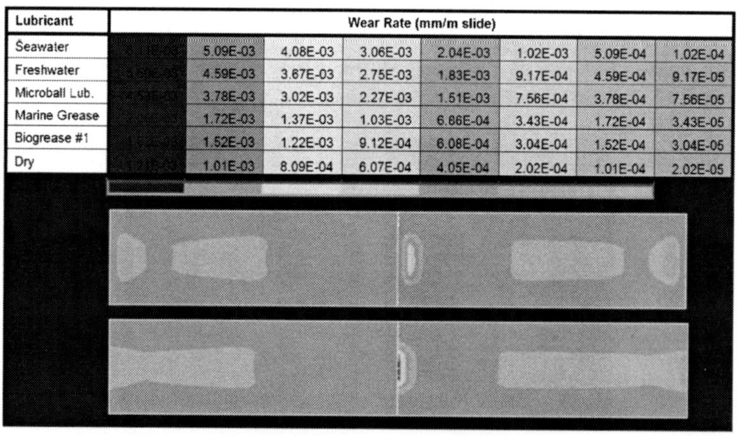

Figure 9: Wear map comparison – chamfer panel (top), Original panel (bottom).

2.3.2 Solution tested

The chamfered panel is tested using the FEA model developed for the original panel alignment modelling and is found to significantly improve the panel's performance. Chamfer optimisation tests indicate that a chamfer geometry of 5 by 60mm will have the most significant effect with stress reductions of around 70% and wear and friction reductions of around 60% compared with the original panel for all offsets up to 5mm in height.

3 Discussion

The multidisciplinary methodology adopted in this case is shown to be necessary to provide a sufficiently broad understanding of the slipway contact case. It is shown that the tribometer testing alone does not give a full enough understanding of the real world issues affecting slipway friction and wear, and that with the large scale test rigs required to investigate panel misalignment effects experimentally proving impractical, it is appropriate to continue the research using finite element simulations of the contact informed by the previous tribometer testing results.

The initial tribometer testing give valuable insight into the role of lubricants in the lifeboat slipway case, it is found that the composite selected is unsuitable for dry running due to its high friction coefficient and must therefore be used with lubricants. It is also shown that the presence of lubricants in the contact disrupts the formation of a protective graphite transfer layer to mitigate wear and friction effects, this indicates that the inclusion of graphite in the currently selected composite is unnecessary, and that other similar composite materials without included graphite may prove more suitable.

With the use of lubricants shown to be necessary the tribometer tests indicate that all lubricants tested could potentially reduce the friction to operable levels, with water lubrication particularly interesting due to its low cost and environmental impact.

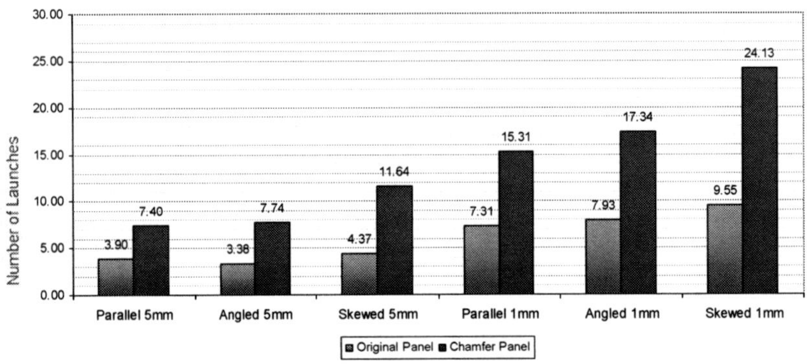

Figure 10: Launches/recoveries required for the slipway panel to fail for original and modified slipway panels using freshwater lubrication.

3.1 Panel lifespan

The Finite Element results indicate that the panel misalignments observed during real world slipway surveys are an important contributor to the apparent friction along the slipway, but it is shown that these effects can be significantly reduced by incorporating a chamfer into the panel design. The use of chamfered panel to reduce misalignment effects should allow lifeboat slipways to operate reliably using water lubrication, and is also predicted to significantly improve the expected lifespan of the composite panels.

3.2 Cost

The use of water lubrication in conjunction with the chamfered panels is shown using life cycle analysis to offer significant financial savings compared with the current manually applied grease lubrication as well as better environmental and safety performance. It is thought that the adoption of either sea or freshwater lubrication systems and chamfer panels will save almost £200k/p.a. compared to

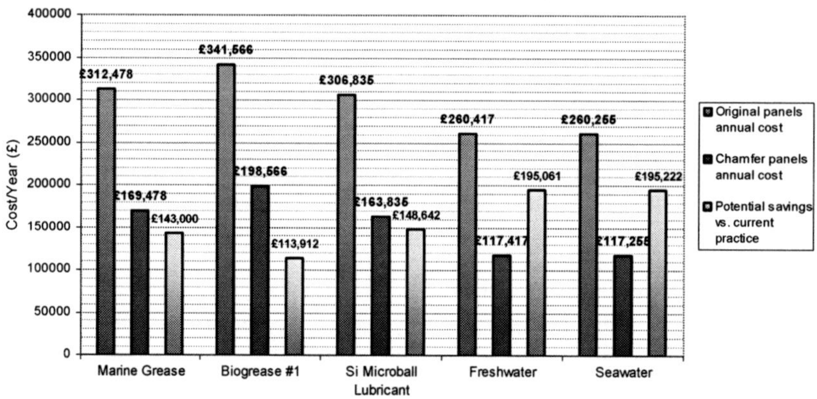

Figure 11: Slipway operating costs/year – original panel vs. chamfer panel.

current practice, a saving of 62%. This is the combined result of reduced panel replacement costs due to the reduction in panel misalignment edge effects using the chamfered panels, and the reduced cost of the lubricant itself.

Table 2: Panel misalignment effects on friction coefficients.

Offset dist. (mm)	Parallel Misalignment		Angled Misalignment		Skewed Misalignment	
	Original	Chamfer	Original	Chamfer	Original	Chamfer
0	0.028	0.054	0.028	0.054	0.028	0.054
1	0.503	0.109	0.247	0.227	0.164	0.107
2	1.035	0.210	0.492	0.464	0.368	0.265
3	1.565	0.313	0.742	0.520	0.582	0.385
4	1.752	0.417	0.984	0.662	0.827	0.606
5	2.194	0.519	1.215	1.123	1.042	0.806

3.3 Friction performance

The FEA simulations show that the use of a chamfer on the slipway panels can greatly reduce the friction contributions of panel misalignments and this is shown in table 2 for a typical 1mm offset.

4 Conclusions

The research conclusions can be summarised as follows:

- It is shown that consistent, low friction using composite slipway panels is only reliably achievable using lubricants.
- The use of biogrease #1 is shown to be as effective as current marine grease practice and this can be considered a candidate for direct substitution with due consideration given to the increased cost of biogrease #1.
- The use of water lubrication is shown to be effective under good slipway alignment conditions and is also shown to offer significant environmental and economic benefits.
- Current slipway winch specifications are shown to be the limiting factor when calculating friction specifications on 1 in 5 slipways and these should be increased as described above.
- The effects of contaminated contact conditions (i.e. windblown sand) are shown to be severe, however real world instances of this were rare and this can be discounted as a systemic factor in slipway panel wear.
- Slipway panel geometry and alignment is shown to be an important factor in determining the real world friction and wear performance of composite slipway panels.
- The use of a modified chamfer panel in place of the current panel design is shown to reduce the effects of panel misalignments leading to reduced wear and friction on slipway panels.
- Slipway panels are shown to fail due to friction specifications rather that panel material failure so that slipway wear performance can be assessed at each slipway station using winch loads criteria.
- Slipway panel failure criteria can be developed by assessing the bearing area of the panel surface.
- Reducing panel misalignment and using chamfered slipway panels is shown to more than double current panel life spans.
- Using water lubrication and chamfered slipway panels could potentially save up to £195k per year whilst improving the safety, reliability and panel lifespan of slipway launched lifeboat operation.

5 Recommendations

Recommendations for slipway lifeboat operation derived from this research are shown below:

- Winch specifications for various slipway angles should be assessed using calculations outlined above. This shows that increasing the winch line pull specification from 12 tonnes to 13.7 tonnes for 1 in 5 gradient slipways will allow for a far greater friction coefficient tolerance. By the same analysis, for shallower 1 in 6 slipways the winch specification can be relaxed to 11.5 tonnes.
- It is recommended to adopt water lubrication systems across the RNLI slipway station network as this is shown to be effective and reliable while also reducing operating costs.
- It is recommended that the use of greases along the slipway be relegated to a backup system for water lubrication due to the economic, environmental and safety implications inherent in direct grease application to the slipway.
- The direct substitution of marine grease with biogrease #1 is shown to be feasible and this should be considered with appropriate assessment of the relative importance of economic and environmental factors.
- Panel misalignments are shown to be a significant factor in real world slipway friction and a maximum panel alignment specification of 2mm is recommended to prevent high friction and wear occurrences. This target should be feasible within current slipway panel fitting practices.
- Slipway condition monitoring using winch loading is shown to be effective at assessing slipway performance and it is recommended that this practice be adopted across all slipway stations.
- Individual slipway panel conditions can be assessed using the bearing area failure criteria developed, and this is recommended for adoption in conjunction with winch load monitoring.
- Further testing of alternative materials is proposed as the inclusion of graphite in the current composite material is shown to be unnecessary and more suitable materials may be available.

References

[1] Warner, O: *'The lifeboat service: a history of the Royal National Lifeboat Institution, 1824-1974'* London : Cassell, (1974)

[2] Leach, N: *'A Century of RNLI Motor Lifeboats'* Landmark Publishing Ltd. (2007)

[3] Hudson, F, D: *'The design and development of modern SAR craft - A personal view'* The Royal Institution of Naval Architects, Drydocks, Launching, and Shiplift Conference, pp79-82. (2003)

[4] Austen, S: RNLI Head of Engineering Support - Private communication, 5th June (2008)

[5] Thomas, B, Hadfield, M, Austen, S: *'Wear and Friction Modelling on Lifeboat Launch Systems'* Tribology Transaction, in press (2009)

[6] Thomas, B, Hadfield, M, Austen, S: *'Experimental Wear Modelling of Lifeboat Slipway Launches'* Tribology International, Vol. 42, Issues 11-12, p1706-14 (2009)

[7] Thomas, B, Hadfield, M, Austen, S: *'Wear Observations Applied to Lifeboat Slipway Launches'* WEAR, Vol. 267, Issue 11, p2062-9 (2009)

[8] Sinha, Sujeet, K: *'Wear Failures of Plastics'* p1019-27, in *'ASM Handbook: Failure Analysis and Prevention Vol. 11: Failure Analysis and Prevention'*, ASM International (2002)

Static load distribution in ball bearings including the effects of temperature and fit

M. Ricci
Space Mechanics and Control Division, INPE, Brazil

Abstract

A numerical procedure for internal loading distribution computation in statically loaded, single-row, angular-contact ball bearings, subjected to a known combined radial and thrust load, which must be applied so that to avoid tilting between inner and outer rings, is used to find the load distribution differences between a loaded unfitted bearing at room temperature, and the same loaded bearing with interference fits, such might experience radial temperature gradients between inner and outer rings. For each step of the procedure it is required the iterative solution of $Z + 2$ simultaneous non-linear equations – where Z is the number of the balls – to yield exact solution for axial and radial deflections, and contact angles.
Keywords: ball, bearing, static, load, numerical, method, temperature, fit.

1 Introduction

Ball and roller bearings, generically called *rolling bearings*, are commonly used machine elements. They are employed to permit rotary motions of, or about, shafts in simple commercial devices such as bicycles, roller skates, and electric motors. They are also used in complex engineering mechanisms such as aircraft gas turbines, rolling mils, dental drills, gyroscopes, and power transmissions.

This work is devoted to study of the internal loading distribution in statically loaded single-row angular-contact ball bearings. Several researchers have studied the subject as, for example, Stribeck [1], Sjoväll [2], Jones [3] and Rumbarger [4]. The methods developed by them to calculate distribution of load among the balls and rollers of rolling bearings can be used in most bearing applications because rotational speeds are usually slow to moderate. Under these speed conditions, the effects of rolling element centrifugal forces and gyroscopic

WIT Transactions on Engineering Sciences, Vol 66, © 2010 WIT Press
www.witpress.com, ISSN 1743-3533 (on-line)
doi:10.2495/TD100191

moments are negligible. At high speeds of rotation these body forces become significant, tending to alter contact angles and clearance. Thus, they can affect the static load distribution to a great extension.

Harris [5] described methods for internal loading distribution in statically loaded bearings addressing pure radial; pure thrust (centric and eccentric loads); combined radial and thrust load, which uses radial and thrust integrals introduced by Sjoväll; and for ball bearings under combined radial, thrust, and moment load, initially due to Jones.

There are many works describing the parameters variation models under static loads but few demonstrate such variations in practice, even under simple static loadings. The author believes that the lack of practical examples is mainly due to the inherent difficulties of the numerical procedures that, in general, deal with the resolution of several non-linear algebraic equations that must to be solved simultaneously.

In an attempt to cover this gap studies are being developed in parallel [6–10]. In this work a numerical procedure, described in [7], for internal load distribution computation in statically loaded, single-row, angular-contact ball bearings subjected to a known external combined radial and thrust load, so that no tilt is allowed between inner and outer rings, is used to find the load distribution differences between a loaded bearing with clearance fits at room temperature, and the same loaded bearing with interference fits, such might experience radial temperature gradients between inner and outer rings.

In the most usual situation, angular contact bearings would first be fitted, with interference or clearance defined at room temperature, to their respective shaft and housing; then a defined axial "hard" preload would be applied and subsequently in operation the bearings might experience radial temperature gradients between inner and outer rings.

Generally, ball bearings and other radial rolling bearings such as cylindrical roller bearings are designed to have a diametral clearance in the no-load state. Due to this radial clearance the bearing also can experience an axial play. Removal of this axial freedom causes the ball-raceway contact line to assume an oblique angle with respect to the radial plane; hence, a contact angle different from zero will occur. This angle is called *free contact angle* and is a function of clearance built into the unloaded bearing and the raceway groove curvatures.

Press or shrink fitting of the inner ring on the shaft causes the inner ring to expand slightly. Similarly, press fitting of the outer ring in the housing causes the former member to shrink slightly. Thus, the bearing's diametral clearance will tend to decrease.

Thermal conditions of bearing operation can also affect the diametral clearance. Heat generated by friction causes internal temperatures to rise. This in turn causes expansion of the shaft, housing, and bearing components. Depending on the shaft and housing materials and on the magnitude of thermal gradients across the bearing and these supporting structures, clearance can tend to increase or decrease. It is also apparent that the thermal environment in which a bearing operates may have a significant effect on clearance.

2 Static load distribution under combined radial and thrust load in ball bearings

Having defined in other works analytical expressions for geometry of bearings and for contact stress and deformations for a given ball or roller-raceway contact (point or line loading) in terms of load, it is possible to consider how the bearing load is distributed among the rolling elements. In this section a specific load distribution consisting of a combined radial and thrust load, which must be applied to the inner ring of a statically loaded ball bearing, so that no tilt is allowed between inner and outer rings, is given.

Let a ball bearing with a number of balls, Z, symmetrically distributed about a pitch circle according to Fig. 1, to be subjected to a combined radial and thrust load, so that a *relative axial displacement*, δ_a, and a *relative radial displacement*, δ_r, between the inner and outer ring raceways may be expected. Let $\psi = 0$ to be the angular position of the maximum loaded ball.

Figure 2 shows the initial and final curvature centers positions at angular position ψ, before and after loading, considering the centers of curvature of the raceway grooves fixed with respect to the corresponding raceway. If δ_a and δ_r are known, the contact angle at angular position ψ, after the combined load has been applied, is given by

$$\beta(\psi) = \cos^{-1}\left(\frac{A\cos\beta_f + \delta_r\cos\psi}{A+\delta_n}\right), \tag{1}$$

where A is the distance between raceway groove curvature centers for the unloaded bearing, β_f is the free-contact angle and δ_n is the total normal deflection at the contacts.

Also, from Fig. 2,

$$\delta_a = (A+\delta_n)\sin\beta - A\sin\beta_f, \tag{2}$$

and we can arrive in the expression for the extend of the *loading zone*, that is given by

$$\psi_l = \cos^{-1}\left\{\frac{A}{\delta_r}\left[\cos\left(\sin^{-1}\left(\frac{\delta_a + A\sin\beta_f}{A}\right)\right) - \cos\beta_f\right]\right\}. \tag{3}$$

From (1), the total normal approach between two raceways at angular position ψ, after the combined load has been applied, can be written as

$$\delta_n(\psi) = A\left(\frac{\cos\beta_f}{\cos\beta} - 1\right) + \frac{\delta_r\cos\psi}{\cos\beta}. \tag{4}$$

From Fig. 2 and (4) it can be determined that s, the distance between the centers of the curvature of the inner and outer ring raceway grooves at any rolling element position ψ, is given by

$$s(\psi) = A + \delta_n = A\frac{\cos\beta_f}{\cos\beta} + \frac{\delta_r\cos\psi}{\cos\beta}. \tag{5}$$

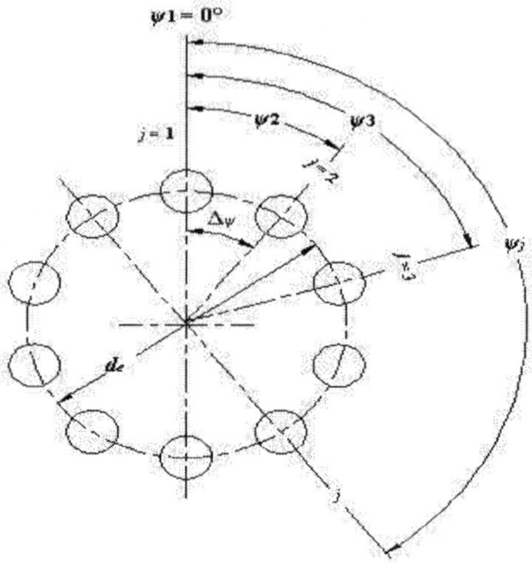

Figure 1: Ball angular positions in the radial plane that is perpendicular to the bearing's axis of rotation, $\Delta\psi = 2\pi/Z$, $\psi_j = 2\pi/Z(j-1)$.

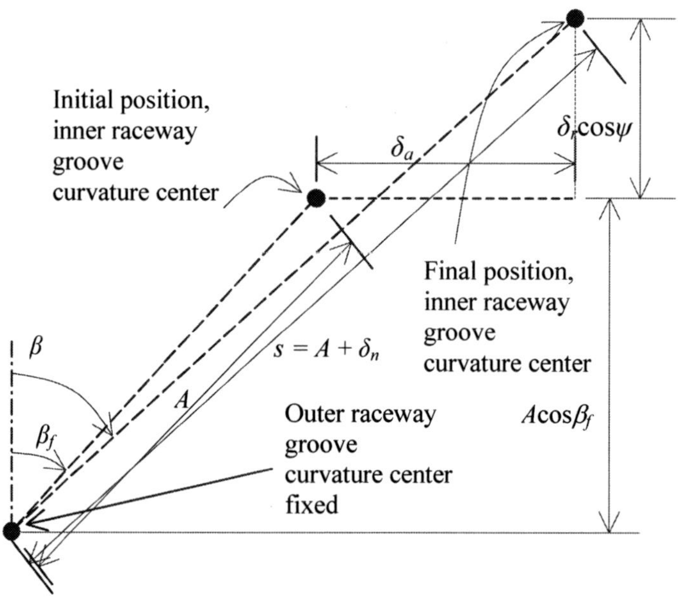

Figure 2: Initial and final curvature centers positions at angular position ψ, with and without applied load.

From (2) and (5) yields, for $\psi = \psi_j$,

$$\delta_a - \delta_r \tan\beta_j \cos\psi_j - A\frac{\sin(\beta_j - \beta_f)}{\cos\beta_j} = 0, \quad j = 1,\ldots,Z. \tag{6}$$

From load-deflection relationship for ball bearings and (4) yields, for $\psi = \psi_j$,

$$Q_j = K_{nj}\left[A\left(\frac{\cos\beta_f}{\cos\beta_j} - 1\right) + \frac{\delta_r \cos\psi_j}{\cos\beta_j}\right]^{3/2}, \quad j = 1,\ldots,Z. \tag{7}$$

If a thrust load, F_a, and a radial load, F_r, are applied then, for static equilibrium to exist

$$F_a = \sum_{j=1}^{Z} Q_j \sin\beta_j, \tag{8}$$

$$F_r = \sum_{j=1}^{Z} Q_j \cos\beta_j \cos\psi_j. \tag{9}$$

Additionally, each of the normal ball load components produces a moment about of the inner ring center of mass, in the plan, which passes through the bearing rotation axis and contains the external radial load (moments about the other two perpendicular plans are self-equilibrating). For static equilibrium, the thrust load, F_a, and/or the radial load, F_r, must exert a moment, M, about of the inner ring center of mass, which must be equal the sum of the moments of each rolling element load, that is,

$$M = -\sum_{j=1}^{Z} Q_j \sin\beta_j\left[(R_i + \delta_r \cos\psi_j)\cos\psi_j - \delta_r\right], \tag{10}$$

where

$$R_i = d_e/2 + (f_i - 0.5)D\cos\beta_f$$

expresses the locus of the centers of the inner ring raceway groove curvature radii.

Substitution of (7) into (8) yields

$$F_a - \sum_{j=1}^{Z} K_{nj} \sin\beta_j\left(A\left(\frac{\cos\beta_f}{\cos\beta_j} - 1\right) + \frac{\delta_r \cos\psi_j}{\cos\beta_j}\right)^{3/2} = 0. \tag{11}$$

Similarly,

$$F_r - \sum_{j=1}^{Z} K_{nj} \cos\psi_j \cos\beta_j\left(A\left(\frac{\cos\beta_f}{\cos\beta_j} - 1\right) + \frac{\delta_r \cos\psi_j}{\cos\beta_j}\right)^{3/2} = 0. \tag{12}$$

Equations (6), (11) and (12) are $Z + 2$ simultaneous non-linear equations with unknowns δ_a, δ_r, and β_j, $j = 1,\ldots, Z$. Since K_{nj} are functions of final contact angle, β_j, the equations must be solved iteratively to yield an exact solution for δ_a, δ_r and β_j.

3 Numerical results

To show an application of the theory developed in this work a numerical example is presented, which uses The Newton–Rhapson method to solve the simultaneous nonlinear equations (6), (11) and (12). To take into account the effects of temperature and fitting I have considered the principal relationships between interference fittings, thermal gradients, surface finish and changes in diametral clearance, as described in [5].

I have chosen the 218 angular-contact ball bearing as example, which was also used by [5]. The 218 angular-contact ball bearing has a 0.09 m bore, a 0.16 m o.d. and is manufactured to ABEC 7 tolerance limits. The bearing is mounted on a hollow steel shaft of 0.0635 m bore with a k6 fit and in a titanium housing having a effective o.d. of 0.2032 m with an M6 fit. Considers that the inner ring operates at a mean temperature of 148.9°C, that the outer ring is at 121.1°C and that the bearing was assembled at 21.1°C.

There are three steps in the numerical procedure. The first, considering the bearing unfitted at assembling temperature; the second, considering the fits above at assembling temperature; and the third, considering the fits above at operational temperatures for the inner and outer rings. Before each step the geometry of the bearing is obtained from which, the nonlinear equations are solved simultaneously to obtain radial and axial deflections and contact angles.

Figs. 3 to 12 show some parameters, as functions of the applied thrust load, for the three steps of the procedure and for some values of the applied radial load.

Figs. 3 and 4 show the normal ball loads for the maximum and minimum loaded ball, respectively. There is a better loading distribution with the increase of thrust load; and there are slight decreases (increases) in normal ball load, for the maximum (minimum) loaded ball, when it passes from first to second, and

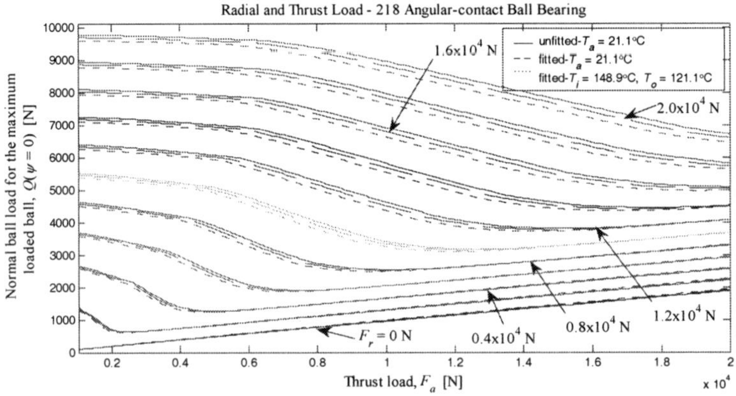

Figure 3: Normal ball load for the maximum loaded ball, $Q(\psi = 0)$, as a function of the thrust load, F_a.

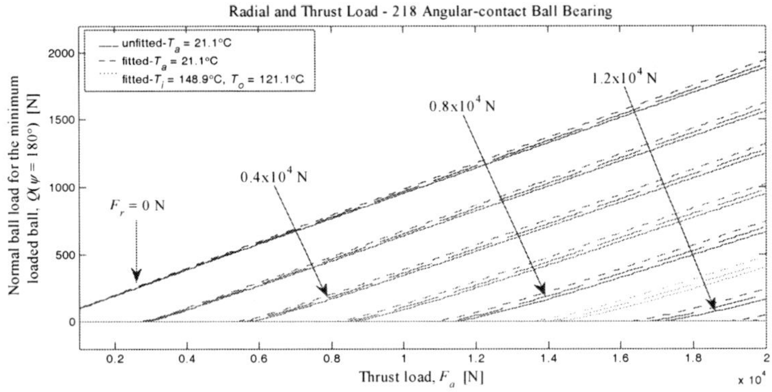

Figure 4: Normal ball load for the minimum loaded ball, $Q(\psi = 180°)$, as a function of the thrust load, F_a.

from second to third steps. Similar behavior is also observed for other parameters and will not be mentioned here. For this loading range, the ball at angular position $\psi = 0$ is always loaded. This is not the case for the ball at angular position $\psi = 180°$. For zero applied radial load the normal ball load is the same for the maximum and minimum loaded ball.

Figs. 5 and 6 show the contact angle for the maximum and minimum loaded ball, respectively. The straight lines represent the free contact angles for the three steps of the procedure. For zero applied radial load the contact angle is always greater than the free contact angle. In this case, the bearing is under thrust loading and all balls have the same load and contact angles. If the contact angle falls below the free contact angle the balls are unloaded. With the increase in applied radial load the free contact angle ceases to be a reference for unload. In this case, the contact angle may drop to values much lower than the free contact angle value - as low as 3° - for a loaded ball, or may rise to values greater than the free contact angle value, for an unloaded ball.

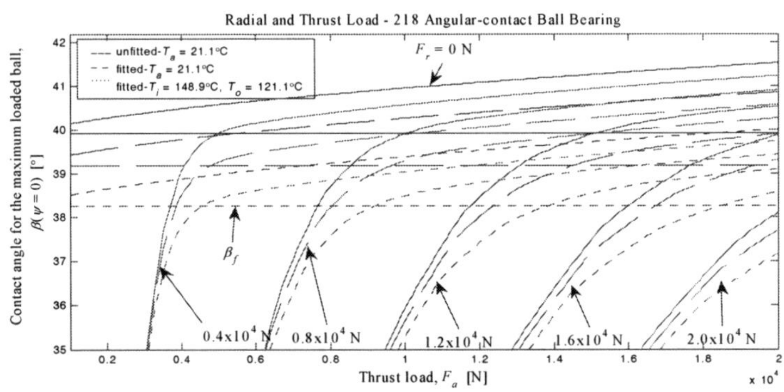

Figure 5: Contact angle for the maximum loaded ball, $\beta(\psi = 0)$, as a function of the thrust load, F_a.

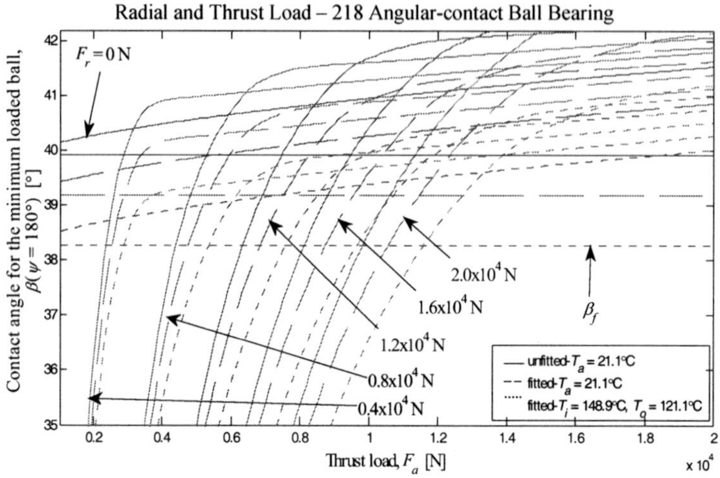

Figure 6: Contact angle for the minimum loaded ball, $\beta(\psi = 180^{o})$, as a function of the thrust load, F_a.

Figs. 7 and 8 show the distance between the curvature centers for the maximum and minimum loaded ball, *s*, respectively. The straight lines represent the curvature centers distance in the shifted (unload) position, *A*. Simply remove *A* from *s* for total deformation along the contact line. If *s* falls below *A* is an indication of unloading.

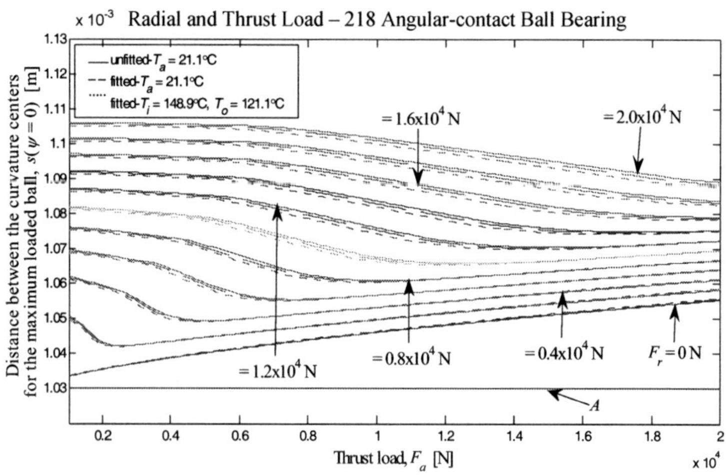

Figure 7: Distance between the curvature centers for the maximum loaded ball, $s(\psi = 0)$, as a function of the thrust load, F_a.

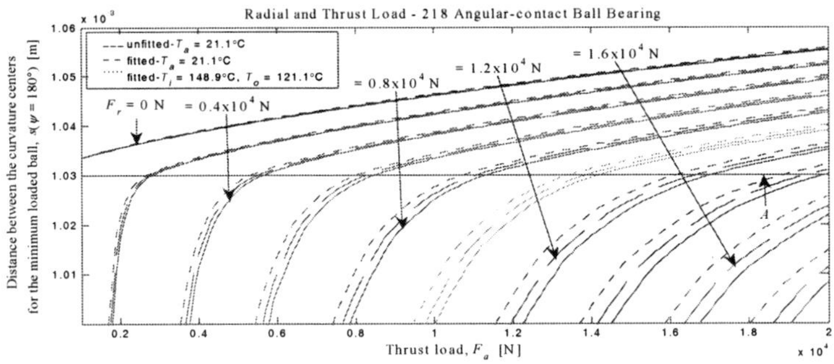

Figure 8: Distance between the curvature centers for the minimum loaded ball, $s(\psi = 180^o)$, as a function of the thrust load, F_a.

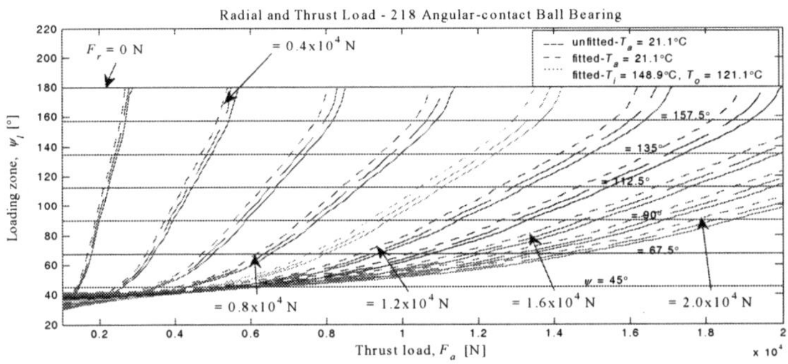

Figure 9: Loading zone, ψ_l, as a function of the thrust load, F_a.

Fig. 9 shows the loading zone. The straight lines represent the angular ball positions. Under radial load, increasing the thrust load there is an increase of the loading angle, whose maximum value is 180°.

Figs. 10 and 11 show the axial and radial deflections, respectively.

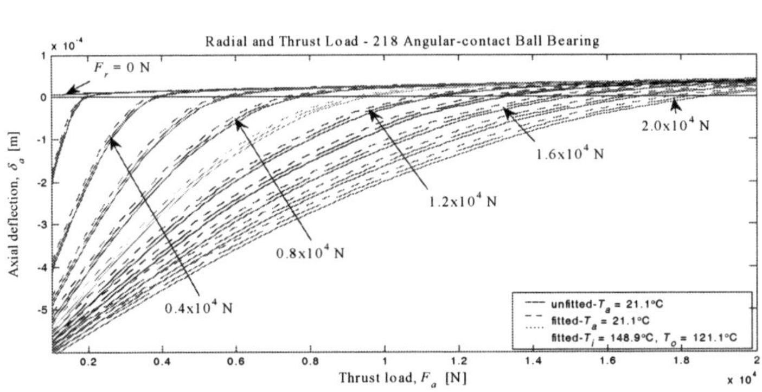

Figure 10: Axial deflection, δ_a, as a function of the thrust load, F_a.

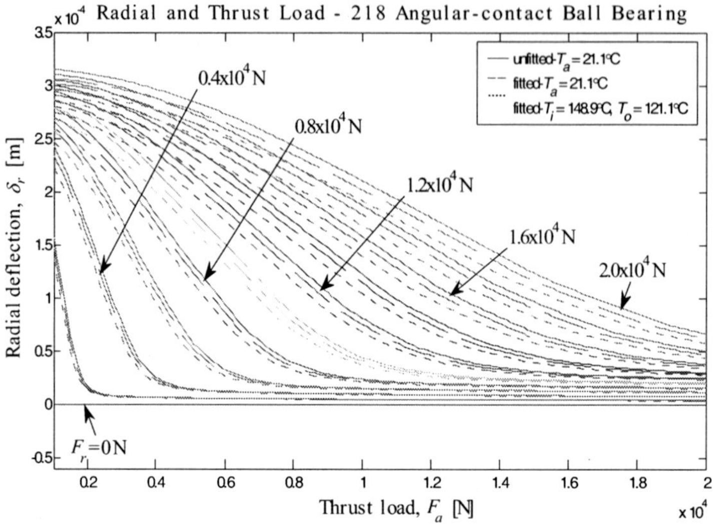

Figure 11: Radial deflection, δ_r, as a function of the thrust load, F_a.

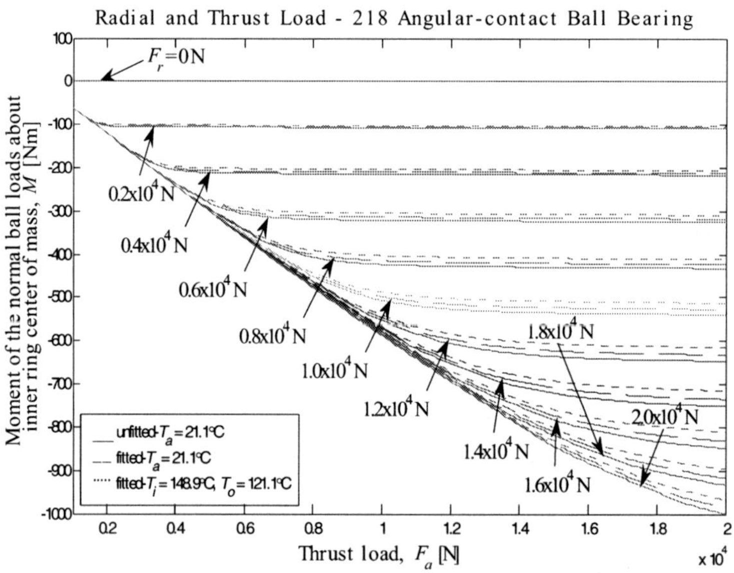

Figure 12: Moment of the normal ball loads about inner ring center of mass, M, as a function of the thrust load, F_a.

Fig. 12 shows the moment, M, such must be exerted by the thrust load, F_a, and/or the radial load, F_r, about of the inner ring center of mass, in order of prevent the angular displacement of the inner ring with respect the outer ring.

4 Conclusion

This work used a procedure for get numerically, accurately and quickly, the static load distribution of a ball bearing under axial and radial loading, taking into account the influence of fits and thermal gradients. Precise applications, as for example, space applications, require a precise determination of the static loading. Models available in literature are approximate and often are not compatible with the desired degree of accuracy. This work can be extended to determine the loading on high-speed bearings where centrifugal and gyroscopic forces do not be discarded. The results of this work can be used in the accurate determination of the friction torque of the ball bearings, under any operating condition of temperature and speed.

References

[1] Stribeck, R., Ball Bearings for Various Loads. *Trans. ASME* **29**, pp. 420–463, 1907.

[2] Sjoväll, H., The Load Distribution within Ball and Roller Bearings under Given External Radial and Axial Load. *Teknisk Tidskrift,* Mek., **h.9**, pp. 97–102, 1933. (http://runeberg.org/tektid/1933m/0099 .html)

[3] Jones, A., *Analysis of Stresses and Deflections,* New Departure Engineering Data: Bristol, 1946.

[4] Rumbarger, J., Thrust Bearings with Eccentric Loads. *Machine Design,* Feb. 15, 1962.

[5] Harris, T., *Rolling Bearing Analysis*, 4th ed., John Wiley & Sons Inc.: New York, 2001.

[6] Ricci, M. C., Ball bearings subjected to a variable eccentric thrust load. *DINCON'09 Proceedings of the 8th Brazilian Conference on Dynamics, Control and Applications*, May, 18-22, Bauru, Brazil, 2009. ISBN: 978-85-86883-45-3.

[7] Ricci, M. C., Internal loading distribution in statically loaded ball bearings, *ICCCM09 1st International Conference on Computational Contact Mechanics, Program and Abstracts*, p. 21-22, Sept. 16-18, Lecce, Italy, 2009.

[8] Ricci, M. C., Internal loading distribution in statically loaded ball bearings subjected to a combined radial and thrust load. *6th ICCSM Proceedings of the 6th International Congress of Croatian Society of Mechanics*, Sept. 30 to Oct. 2, Dubrovnik, Croatia, 2009. ISBN 978-953-7539-11-5.

[9] Ricci, M. C., Internal loading distribution in statically loaded ball bearings subjected to a combined radial, thrust, and moment load. *Proceedings of the 60th International Astronautical Congress*, October, 12-16, Daejeon, South Korea, 2009. ISSN 1995-6258.

[10] Ricci, M. C., Internal loading distribution in statically loaded ball bearings subjected to an eccentric thrust load. Accepted to publication at The *Mathematical Problems in Engineering*, 2009.

[11] Ricci, M. C., Internal loading distribution in statically loaded ball bearings subjected to a combined radial, thrust, and moment load, including the effects of temperature and fit. Presented at *11th Pan-American Congress of Applied Mechanics*, January, 04-10, Foz do Iguaçu, Brazil, 2010.

[12] Hamrock, B. J. and Anderson, W. J., *Arched-Outer-Race Ball-Bearing Considering Centrifugal Forces*. NASA TN D-6765, 1972.

[13] Hertz, H., On the contact of Rigid Elastic Solids and on Hardness. in *Miscellaneous Papers*, MacMillan, London. 163-183, 1896.

[14] Hamrock, B. J. and Anderson, W. J., *Rolling-Element Bearings*. NASA RP 1105, 1983.

Section 5
Surface measurements and lubrication

A method for matching fractured surfaces using shadow profilometry

N. H. Maerz[1] & M. C. Hilgers[2]
[1]*Rock Mechanics and Explosives Research Center, Missouri University of Science and Technology, USA*
[2]*Department of Information Science and Technology, Missouri University of Science and Technology, USA*

Abstract

Characterization of fractured surfaces is of interest to forensic scientists because the measurements can be used to investigate if two fractured surfaces have originated from a single common item, thus allowing the scientists to reconstruct shattered objects and structures. Fracture surfaces of object fragments that failed in tension under load or were torn apart because of explosive forces will have essentially matching surfaces. This paper presents an inexpensive technique for constructing a digital image of a three dimensional surface via two dimensional slices of very small objects using shadow profilometry. This method preserves minor details so that measurements can be made to characterize each surface and calculate the likelihood that the two surfaces were disjointed. Samples were created by using a load frame to pull apart round, square, and rectangular rods of various materials, such as high and low carbon steel, aluminium, brass, and copper. Other samples were created by blasting. Various methods and algorithms utilizing optical microscopy, shadow profilometry, and digital image processing were developed to characterize two fracture surfaces to determine if they originated from a single common object.
Keywords: surface roughness, surface characterization, shadow profilometry, forensic investigations.

www.witpress.com, ISSN 1743-3533 (on-line)
doi:10.2495/TD100201

1 Introduction

1.1 Matching disjointed surfaces

Following disasters caused by explosions or other violent crashes, investigators try to re-assemble original structures from pieces of ripped metal to aid in the determination of the cause of the disaster. Typically there are large numbers of fragments involved in the reconstruction, and not all can be easily fitted together. In many cases it is not possible to determine whether two objects fit together or not. Material failure tends to be somewhat plastic rather than perfectly brittle, making the task more difficult. However, fractured surfaces will match at least on the micro level, in a probabilistic sense.

Surface characterization is inherently 3-dimensional in nature. Intuitively, the identifiable features of a surface include its pits, peaks, scraps and so forth. Characterization of a surface requires measuring its texture or roughness so that its discrete features can be better defined.

1.2 Profilometry methods

Profilometry is a method of using some means to create a cross-section of a surface or surface profile. Mechanical profilometry is a time honoured method of measuring surface profiles, but is increasingly difficult to do on very small surfaces. LADAR is (LAser Detection and Ranging) is new technology that can accurately map surfaces, but requires very expensive equipment, and still has limitations on the size of surfaces that can be scanned [2, 3].

Shadow profilometry is the most efficient and scale independent method available for 3-D characterization [1].

Shadow profilometry is a technique in which an edge of light/shadow is used to trace a surface profile. Multiple profiles produced by moving laterally across a fragment can be used to create a 3-D surface. The principle is that if one creates a plane of light or shadow and intersects it with an undulating surface at an angle of 45°, then the resultant line of intersection follows the topography of the surface and creates a linear profile. To understand this technique better, consider the shadow cast by straight edge on a surface. If the surface is flat then the edge of the shadow will remain straight, but if the surface has an irregularity, such as a divot, the shadow's edge will have a corresponding curve. Deeper divots will result in more pronounced curvature in the shadow.

A demonstration of a shadow edge is given in Fig. 1. Historically, this principle is put forth by the Schmaltz microscope [4]. It has been used in the mining industry to measure the roughness of rock surfaces to predict their frictional resistance to shearing movement [1]. The particular focus of this investigation is to extend this method to a smaller scale (roughness amplitudes of less than 0.1 mm) suitable for forensic analysis.

The same principle can be exploited using laser striping [5, 6]. Multiple laser lines can be used to get better coverage on the surface (Fig. 2). However, laser lines have a finite width, and because of laser speckle, cannot be used at small

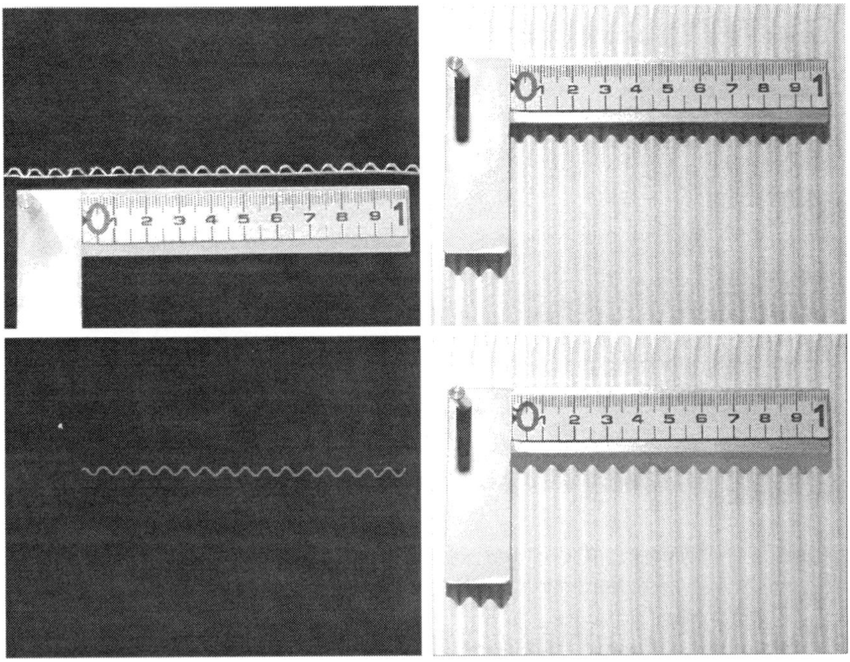

Figure 1: Principle of shadow profilometry highlighting the topography of a corrugated cardboard surface (upper left). Shadow (upper right) is identified (lower right) and the edge is isolated creating a linear profile (lower left).

scales (measuring amplitudes of less than 2 mm). Profiles produced by shadow edges on the other hand enjoy a certain degree of scale invariance and can be used on very small objects.

Although the shadow edge can only reveal one profile at a time, multiple profiles can be generated by moving the shadow across the surface while acquiring successively images. These images can be combined to create a 3-D surface reconstruction and viewed as a wireframe approximation of a surface.

1.3 Surface reconstruction

Surfaces have been reconstructed in other settings from specific shadow data, such as looking at the shapes of shadows cast by the surface [7], or from the movement of shadows as a light follows a known trajectory above the object [8] and from around the object [9]. Research has also gone into reconstructing surfaces from contours in a family of cross-sectional images [10]. These research projects have had a wide range of success in generating the overall shape and large features of an object, have had little to no success in reconstructing the small-scale, fine features (roughness) of these objects. These minor features play a significant role in roughness calculations and thereby in determining whether two pieces fit together.

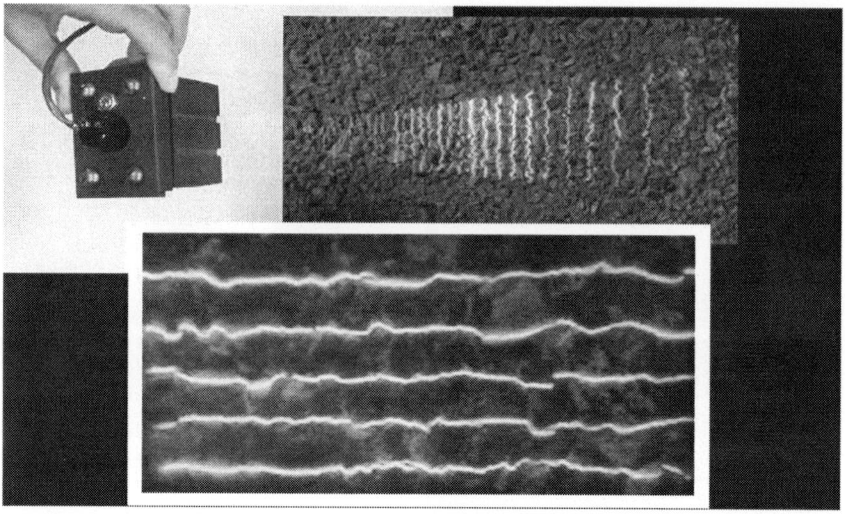

Figure 2: Example of profiling of a prepared concrete surface using laser striping hardware [5, 6].

2 System hardware

This system was developed with inexpensive off-the-shelf hardware, and is shown in Figure 3. A sample holder is used to secure the samples. It is mounted on a linear stage, set on top of a scissor jack for up and down movement. A manually adjustable tilt stage is added to help with levelling the sample. The shadow profile is achieved by using a fibre optic line light guide and a straightedge (black painted razor blade), mounted in a custom-designed bracket (Figure 4). The bracket holds the light at a 45 degree angle, and the shadow edge generator at an appropriate position and orientation. The light guide has a fibre optic tube which goes back to a variable intensity light source. The straightedge must be lowered as closely to the sample surface as possible to create a crisp shadow edge. A light baffle is added to prevent diffuse reflected light from weakening the shadow edge. Figure 5 shows an example of a shadow edge profile, with and without the addition of the light baffle.

For imaging, a microscope, a camera, and a computer are required. The microscope selected is an Infinity Infivar continuous focus system that could resolve down to a field of about 0.2 mm. A Sony XCD-SX900 monochrome camera is used because of it high resolution and square pixels (1280 by 960), its non-interlaced (progressive scan) image, and its digital interface (firewire) that negates the need for digitizing hardware and software. The computer for this system is a Sony Vaio laptop computer. It is used because it has firewire input capability.

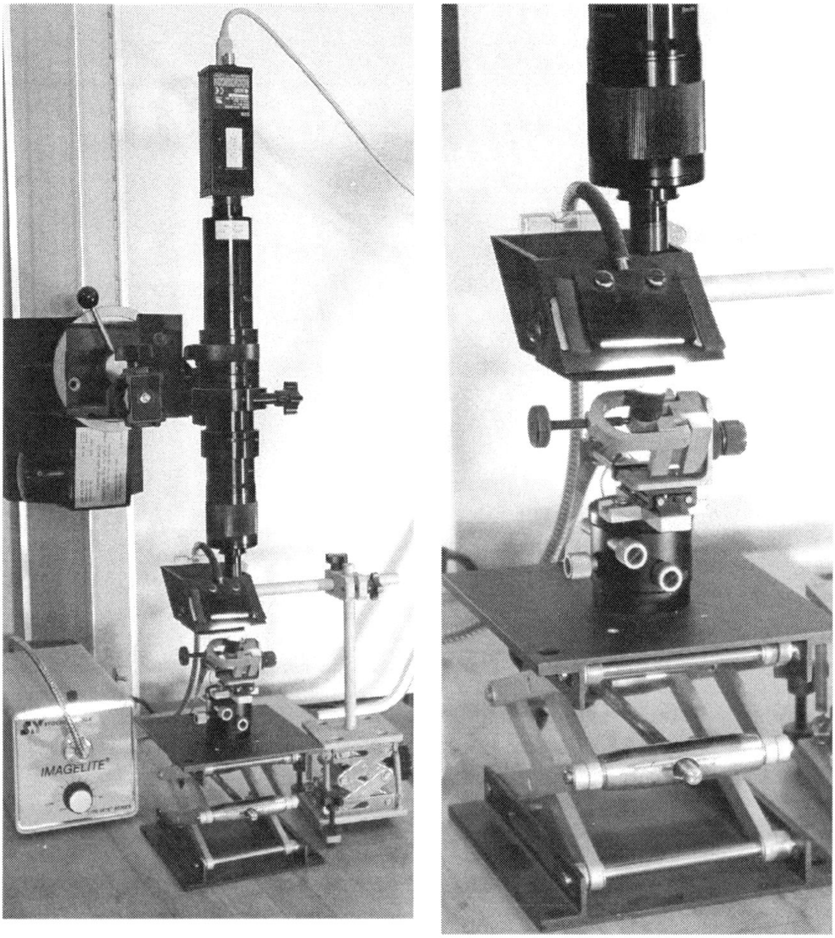

Figure 3: Left: imaging setup (top down) show camera, microscope, light guide, and sample holding equipment with fibre optic light source to the left. Right: (from bottom) scissor lift, tilt stage, linear stage, tilting sample holder, and line light source and guide.

3 System software

Software was developed to move the stage (sample) through the third dimension, acquire the series of images, isolate the shadow edge into a profile, reconstruct the 3-D surface, and make measurements on the profile and surface reconstructions.

The linear stage is controlled by downloading control codes along a standard serial port. Velocity, acceleration, home and limit locations are pre-programmed. The device is first set to the home location. Then an instruction is sent to move to the next position. The device status is queried, and when the

Figure 4: Fibre optic line light guide and shadow edge generator over the sample holder set under the mounting bracket. The microscope objective is visible at the top.

Figure 5: Shadow edge of prepared sample without and with light baffle.

done signal is received, the image can be acquired. This cycle is repeated until the end of the sample is reached.

Images are capture from the Sony camera through a firewire interface at a resolution of 1280 by 960 pixels with a depth of 8 bits per pixel. Images are saved on disk until all images in the set have been acquired. Typically some 50 to 100 profiles are collected for each sample.

Each image is processed in turn to isolate the shadow edge profile (Figure 6). A binary threshold simply scans the image and sets each pixel to white if the pixel's original value is above a previously set threshold level and is set to black otherwise. The second stage is to apply Canny's Edge Detection Algorithm to the resulting image to highlight the regions that change from black to white. Finally, a depth-first-graph-search algorithm is used to discover the shadow's edge. These points, which make up an edge profile, will later be used to calculate roughness statistics and create a digital terrain model of the original sample. Figure 7 shows a series of profiles created form a single sample.

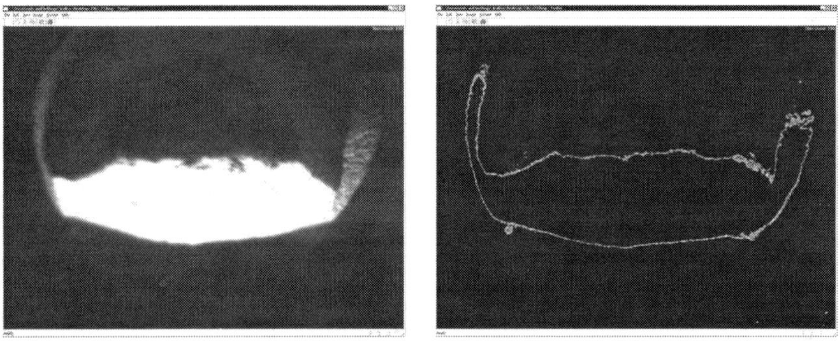

Figure 6: Left: greyscale image of the shadow profile. Right: isolated shadow edge. The lower line (shadow edge) is then pruned to show the actual profile.

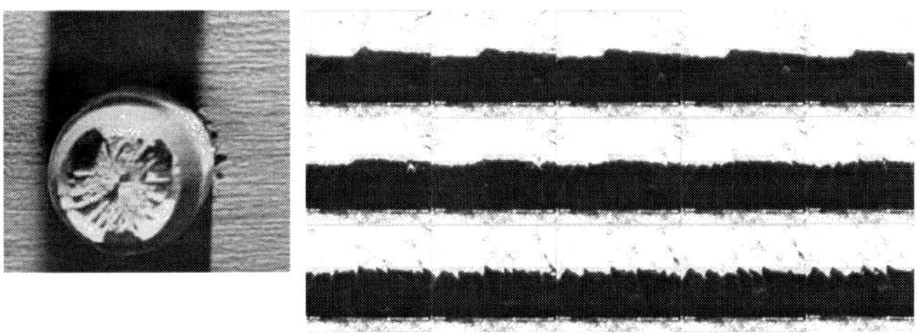

Figure 7: Left: surface sample of hardened steel failed in tension. Right: series of 15 linear profiles across part of the sample.

4 Testing samples

Various different types of samples were used for the purpose of developing and testing this system. In each case matching pairs of samples were used. Rods and bars of various sizes composed of high carbon and mild steel, aluminium, brass, and copper were loaded and failed in tension to create matching surfaces. Blasted samples were created by using steel plates either shattered with point charges or ripped apart using shaped charges. Finally, manufactured surfaces were created by machining one surface and creating a plaster of Paris matching surface.

5 2-D (linear profile) analysis

Linear profile analysis was used for the first level of investigation. This analysis consists of calculating the roughness of each of the series of profiles derived from each sample. Roughness measures, defined elsewhere [1], consist of amplitude measurements such as peak to valley height, levelling depth, root mean and square deviation from centreline, centre line average and others. Slope measures consisted of root mean square of the first derivative, roughness profile index, fractal dimension, and others [1].

Using the multiple profiles of each surface as replicates allows the statistical techniques such as Analysis of Variance (ANOVA) to test whether or not there is a significant difference in the roughness's of two surfaces, i.e. whether the surfaces statistically match or not.

Figure 8 and Table 1 shows the results of tests trying to match surfaces using ANOVA testing on surface roughness of aluminium rods failed in tension. In these tests two pairs of surfaces were tested. Each pair of surfaces (with a common fracture surface) is tested against each other, and against another surface of a sample of the same size and material. In 64 scenarios the correct conclusion was obtained in 62 cases.

6 3-D profile correlation

To align two surfaces (based on topography and features) a surface matching program was created (Figure 9) with an interface to manually align them. Tools were developed to determine the correlation between two surfaces.

Figure 8: Sample G2 matching surfaces, as an example of the samples used in Table 1 (surfaces created in aluminium using static tensile force).

Table 1: Summary of ANOVA testing of 5 control samples pairs (G2 is a sample number and G2 X and G2 Y) indicate opposite sides of the same fracture surface. "Match" indicates that the surfaces have been found to be statistically similar in roughness parameters Ia (average inclination angle), Rp (roughness profile index), Z2 (root mean square of the first derivative), RMS (root mean square), and Rmax (maximum amplitude). A check mark indicates that surfaces were correctly matched or not matched.

Sample pair	Ia	Rp	Z2	RMS
G2 X / G2 Y	Match √	Match √	Match √	Match √
G8 X / G8 Y	Match √	Match √	Match √	Match √
G2 X / G8 Y	No-match √	No-match √	No-match √	No-match √
G8 X / G2 Y	No-match √	No-match √	No-match √	No-match √

Sample pair	Rp	Z2	RMS	Rmax
H4 X / H4 Y	Match √	Match √	Match √	Match √
H5 X / H5 Y	Match √	Match √	Match √	Match √
H4 X / H5 Y	No-match √	No-match √	No-match √	No-match √
H5 X / H4 Y	No-match √	No-match √	No-match √	Match *X*

Sample pair	Rp	Z2	RMS	Rmax
H4 X / H4 Y	Match √	Match √	Match √	No-match *X*
H5 X / H5 Y	Match √	Match √	Match √	Match √
H4 X / H5 Y	No-match √	No-match √	No-match √	No-match √
H5 X / H4 Y	No-match √	No-match √	No-match √	No-match √

Sample pair	Rp	Z2	RMS	Rmax
T3 X, T3 Y	Match √	Match √	Match √	Match √
T4 X, T4 Y	Match √	Match √	Match √	Match √
T3 X, T4 Y	No-match √	No-match √	No-match √	No-match √
T4 X, T3 Y	No-match √	No-match √	No-match √	No-match √

Greyscale elevation maps are created by representing surface elevations by gray tones for each of the two surfaces. The difference map is created when the greyscale elevation maps are compared. Homogenous fields on the difference maps are indicators of good matches, dark and light areas are indications of poor matches. (Fields of really poor matches are highlighted in bright colours.)

Feature maps are used to match surface features. The types of features investigated are peaks, valleys, and ridges. These are identified by processing the greyscale representation of the digital terrain model described previously. Peaks and valleys are easily recognized by searching the terrain model for local minimum and maximum points. The number of peaks and valley points

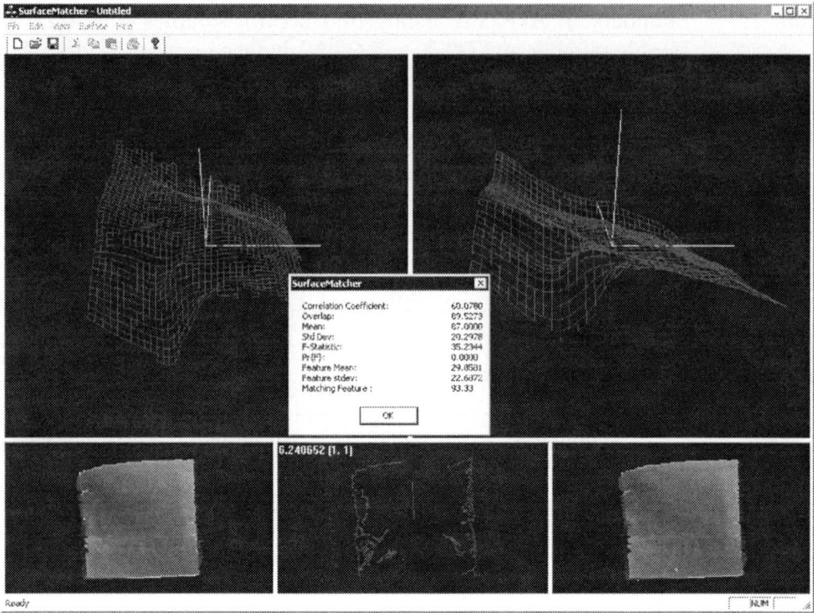

Figure 9: Surface matching program. Two tested surfaces are aligned manually, and then compared automatically. Top left: wireframe image, surface 1. Top right: inverted wireframe image, surface 2. Bottom left: difference map. Bottom centre: feature map (active, allows rotation of one image with respect to the other). Bottom right: difference map highlighting the outliers in yellow. Centre: calculated statistics.

identified is controlled by Gaussian pre-filtering of the image. Figure 10 shows matched opposing opposite surfaces with local minima and maxima flagged.

Within a surface, ridges are areas where the slope drastically changes (downwards to upwards, for example). In image processing, edge detectors are used to segment an image by identifying areas where the difference between pixels is abnormally high. Therefore, using an edge detector to identify ridges in the surface was a natural choice. Specifically, Canny's Edge Detection Algorithm was successfully used to identify ridges in the surface. This is also shown in Figure 10.

Feature correlation is accomplished by analyzing the number of corresponding features between surfaces. Features for each surface are projected onto a single plane, and the features from one surface are matched to the closest feature from the other. The mean and standard deviations of the distances between features is reported, a small mean and standard deviation will correspond to closely matching surfaces.

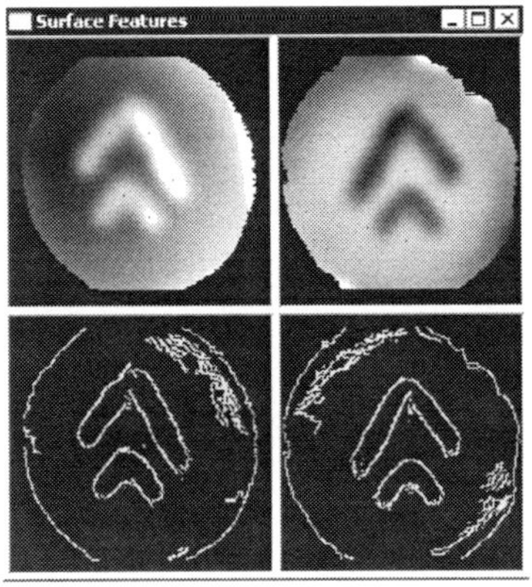

Figure 10: Extracted features from both sides of a manufactured surface. Top: light grey dots are local minima; dark grey dots are local maxima; for the most part the local minima (light grey dots) from one image match the local maxima (dark grey dots) from the other. (White background indicates positive relief, black negative.) Bottom: white lines mark ridges identified by the Canny edge detector.

The Correlation Coefficient is a standard statistical formula used to determine what proportion of the variations in one data set is explained by variations in another. Correlation values near 1 (100%) signify a good match, while values near -1 and 0 signify an inverse match and no relevant match respectively.

The F-Statistic is calculated by assuming one surface is an approximation of the other. Points are randomly selected from the surfaces and are them to calculate the mean-squared error and regression, which are then used to calculate the F-statistic for the model. This statistic can be used to determine the probability of observing two surfaces that match this closely by looking up the corresponding value from an F distribution with 20 numerator degrees of freedom and one denominator degrees of freedom.

7 Summary and conclusions

The methods described in this paper give forensic investigators a novel and unique tool to reconstruct fractured objects by determining if two fractured surfaces are identical mirror images. Shadow profilometry can accurately characterize surfaces 3-D using relatively inexpensive equipment.

Linear profiles can be used by measuring surface roughness. Multiple profiles can be used as replicates, allowing statistical testing to determine if the roughness of the two surfaces in question is the same.

A three dimensional surface, modelled as a wireframe, can be further used to determine the compatibility of two surface maps. Two scanned surfaces can be both compared visually, and surface matching statistics can be generated, while rotating the wireframes to optimize alignment of features.

Acknowledgements

This work was supported by Federal Bureau of Investigation Laboratory Contract No. J-FBI-02-135, Disjointed Fractured Surface Characterization Using Digital Image. Special thanks to Dr. Robert Koons of the FBI for his guidance on this project. Thanks to Christopher P Walker, Dr. Jeff Thomas, Dr. Greg Galecki, Abhay Sonawane, Jordan Gittemeier, and Jessica Shaffer, and Michael Hellrich for their work on this project.

References

[1] Maerz, N. H., Franklin, J. A, and Bennett, C. A., 1990. "Joint Roughness Measurement Using Shadow Profilometry." Int. J. of Rock Mech. Min. Sci. & Geomech. Abst, 27, pp. 329-343.

[2] Kim, H., Haas, C. T., Rauch, A. F., and Browne, C., 2002. "Dimensional Ratios for Stone Aggregates from Three-Dimensional Laser Scans."

[3] Agapakins, J., 1990. "Approaches for Recognition and Interpretation of Workpiece Surface Features Using Structured Lighting". The International Journal of Robotics Research, 9 (5).

[4] Schmaltz, G., 1936. "Technishe Oberflächenkunde." Springer, Berlin, 286 pp.

[5] Maerz, N. H., Chepur, P, Myers, J., J., and Linz, J., 2001a. "Concrete Roughness Measurement Using Laser Profilometry for Fiber Reinforced Polymer Sheet Application." Transportation Research Record, 1775, pp. 132-139.

[6] Maerz, N. H., Nanni, A., Myers, J. J., and Galecki, G., 2001b. "Laser Profilometry for Concrete Substrate Characterization Prior To FRP Laminate Application." Concrete Repair Bulletin, May-Jun. 2001, pp. 4-8.

[7] Daum, M, and Dudek, G., 1998. "On 3-D Surface Reconstruction Using Shape from Shadows." Computer Vision and Pattern Recognition.

[8] Raviv, D., Pao, Y, and Loparo, K., 1989. "Reconstruction of Three Dimensional Surfaces from Two-Dimensional Binary Images." IEEE Transactions on Robotics and Automation, 5 (5).

[9] Seng, T., Len, G., and Chan, K., 1993. "Shape Reconstruction from Shadow and Shading." Industrial Electronics, Control, and Instrumentation.

[10] Ahmed, M., Hemayed, E., Jansing, D., and Farag, A., 1996. "Surface Reconstruction from Serial Cross Sections". Aerospace Applications Conference, 4, pp. 3-10.

The minimum quantity of lubricant (MQL) technique in the surface grinding of steel

E. C. Bianchi[1], R. S. Destro[1], M. H. Oikawa[1], R. M. Sousa[1],
R. C. Canarim[1], R. F. Damasceno[1], P. R. Aguiar[2]
& M. C. Sampaio Alves[3]
[1]*Department of Mechanical Engineering,
UNESP - São Paulo State University, Brazil*
[2]*Department of Electrical Engineering,
UNESP - São Paulo State University, Brazil.*
[3]*Department of Wood Engineering, UNESP - São Paulo State, Brazil*

Abstract

This research aimed to analyze the viability of the minimum quantity of lubricant (MQL) technique towards different methods of lubri-refrigeration in surface grinding of steel, considering process quality, wheel life and the viability of using cutting fluids. The proposal methods were the conventional (abundant fluid flow), the minimum quantity lubrication (MQL) and the optimized method with Webster nozzle (rounded). This analysis was carried out in equal machining conditions, through the assessment of variables such as grinding force, surface roughness, G ratio (volume of removed material/volume of wheel wear), and microhardness. The results showed the possibility of improvement of the grinding process. Besides, there is the opportunity for production of high quality workpieces with lower costs. The MQL technique showed efficiency in machining with lower depths of cut. The optimized method with Webster nozzle applies the fluid in a rational way, without considerable waste. Hence, the results show that industry can rationalize and optimize the application of cutting fluids, avoiding inappropriate disposal, inadequate use and consequently environment pollution.
Keywords: grinding, cutting fluids, MQL, optimization.

WIT Transactions on Engineering Sciences, Vol 66, © 2010 WIT Press
www.witpress.com, ISSN 1743-3533 (on-line)
doi:10.2495/TD100211

1 Introduction

Grinding is a very complex machining process, with many characteristic parameters influencing each other [1]. Its main characteristic is the material removal by the contact between a tool containing abrasive grains (grinding wheel) and the workpiece surface. Each grain of the grinding wheel topographic structure removes a small quantity of material, providing the work surface with a better finishing [2].

According to Hassui and Diniz [3], grinding is the ideal process for a workpiece which needs good surface, dimensional and geometric qualities. Due to these requirements, the process is usually one of the last steps in the machining chain. When the workpiece reaches the grinding stage, it already has high aggregated value from the previous procedures. This makes a possible failure during the finishing process with consequent rejection very expensive and undesirable.

Process industries are seeking for higher production, cost reduction, flexibility, better dimensional and surface quality. This is becoming more and more important to these industries in order to keep, or even increase their market shares in global economy [4]. Grinding is a finishing process, characterized by obtaining results such as precise dimensional tolerance and high superficial quality. However, a limiting factor is the excessive heat generation on the cutting interface. This excessive heat can cause thermal damages to the workpiece, undesirable in a finishing process. Thus, to minimize heat generation, fluids are used to cool and lubricate the cutting zone [5].

In grinding processes, the function of cutting fluids is to remove the excessive heat generated during the cutting operation, lubricate the region where the contact occurs, expel the excess of chips formed, keep the grinding wheel clean, and minimize corrosion on the material involved [6].

Nevertheless, cutting fluids are dangerous substances for the environment. Their disposal is considerably expensive and involves hard procedures. Besides, those substances can cause several diseases after prolonged exposures, such as dermatitis, problems in the respiratory and digestive system, and sometimes even cancer [6].

As shown, cutting fluids play an important role in the productive process, and their use results, most of the times, in a longer life cycle of the tool and in better quality of the machined workpiece. However, their application brings along a few issues, such as difficulties and high cost with recycling and disposal, environmental pollution and chronic diseases for the people exposed. Because of this, a lot of research has been done in order to avoid or minimize cutting fluids use in grinding processes, but without losing the benefits these fluids cause, mainly related to the tool life cycle and the workpiece quality [7].

This paper seeks to analyze lubrication-cooling alternative methods to the conventional method, making it possible a more rational use of cutting fluids, without any losses concerning the quality parameters. These methods are, mainly, the minimum quantity lubrication (MQL), and a fluid application optimized method, with a nozzle proposed by Webster [8]. Through experimentation and analysis of obtained data, the purpose is to verify if the MQL technique can be considered as an alternative to fluid application in grinding, keeping the standards for dimensional and surface quality, characteristics of the process studied. Besides, it is proposed to analyze whether the optimized method with Webster [8] nozzle can be an alternative to the conventional method of application in terms of cutting fluid usage.

In this research, the grinding wheel peripheral velocity and the workpiece velocity were kept constant, and the depth of cut varied in 0.02mm, 0.05mm and 0.08mm. This choice was made in order to simulate three different conditions of grinding: a smooth one (0.02mm), typical of finishing; a severe one (0.08mm); and an intermediate one (0.05mm). Hence, the adopted values can summarize the conditions of grinding in industries. These values were obtained by means of experimental testing. Besides that, the grinding machine used allows a better control of the depth of cut than the other parameters.

2 Materials and methods

The equipment set consists of a SulMecânica 1055E surface grinder; a vitrified superabrasive grinding wheel CBN (350mm external diameter, 127mm internal diameter, 20 mm width, with 5 mm thickness of abrasive material), code SNB151Q12VR2; AISI 4340 workpiece, tempered and quenched (54 HRc average hardness), rectangular shape, 100x200x10mm dimensions. The dressing operation was kept constant, using a multigranular dresser, which did not influence the output variables of the process.

The cutting fluid utilized in the conventional and optimized methods was semi-synthetic soluble oil, with concentration of 5% of fluid in the emulsion, as recommended by the manufacturer. In the conventional method, this fluid is plentifully applied on the contact interface, in a flow rate of approximately 27.5l/min. In the optimized method, a pump makes it possible to apply the fluid at a velocity of 32m/s, the same as the peripheral of the grinding wheel. In this case, the pressure used was about 8bar, and flow rate is the same as the one used in the conventional method (27.5l/min).

The MQL technique consists of atomizing small quantities of fluid in a compressed air line. The MQL system is composed of: compressor, pressure regulator, air flow meter, dosing device and nozzle. It allows a fine adjustment of lubricant/air volume independently, by means of a needle type register. The cutting fluid used was biodegradable vegetal oil with additives for extreme pressure, severe operations and antioxidants. In this experiment, the air flow was pressured nearly 8bar, and cutting fluid rate was 100ml/h.

The optimized method with Webster [8] nozzle requires a special design with rounded surfaces, which can reduce turbulence flow of the cutting fluid. This nozzle was manufactured as close as the author indicates in his paper. This method also requires a pump, in order to apply the fluid at the cutting zone.

With the equipment ready, preliminary experiments were carried out, to define grinding parameters. Keeping the workpiece velocity at 0.0033m/s and the grinding wheel peripheral velocity at 32m/s, both constant, the depth of cut was varied in 0.02mm, 0.05mm and 0.08mm.

The end criterion was the specific volume of the material removed, adopted 5000mm^3/mm. Since there were three different depths of cut, the interval to obtain the specific volume was different in each situation.

The tangential cutting force was measured through the determination of electric power consumed by the engine of the grinder spindle, using an electronic module and a board of data gathering, manipulated by a data gathering program, based on the software LabView, from National Instruments®.

Surface roughness data were obtained measuring R_a, parameter with a rugosimeter Taylor Hobson Surtronic^{3+}, at each 1000mm^3/mm of removed material.

The grinding wheel wear measurement was made using specimen made of AISI 1045 steel for impress the wheel. The impression made was measured by a Tesa Micro-Hite 2D Direct Computer Control (DCC) coordinate measuring machine.

Vickers microhardness was also determined. To obtain this measure, a Buehler 1600-6300 meter was used.

For each depth of cut, three different lubrication-cooling methods were tested, the conventional, the MQL and the optimized method application with Webster nozzle. For each method, the experiments were repeated three times. Thus, a total of 27 experiments were carried out.

3 Results and discussions

In this topic, the gathered data was analyzed. It is important to remember that the workpiece and the grinding wheel peripheral velocities were kept constant, only the depth of the cut varied in 0.02mm, 0.05mm and 0.08mm.

3.1 Tangential cutting force

Figure 1 shows the tangential cutting force behaviour along the experiment for the three depths of cut. It can be observed that in the MQL technique at 0.02 and 0.05 mm depth of cut the cutting force is lower than that observed in the conventional and optimized methods. This can be explained by the fact that the MQL, under those conditions, allows a more effective lubrication in the cutting zone, overcoming the air barrier generated by the grinding wheel through its movement. This way, the strength spent on the friction between the workpiece and the tool is minimized. Similar condition is observed in the optimized method, whose behaviour is quite similar to MQL.

At 0.08 mm depth of cut, the force observed in the MQL is the highest. That grinding condition is the most severe among the ones analyzed, which can be confirmed by an increase in the cutting force. The fact that the forces in the MQL are greater refers to the difficulties in ensuring efficient lubrication. Under these more severe conditions of cutting, high stresses are present. Besides the

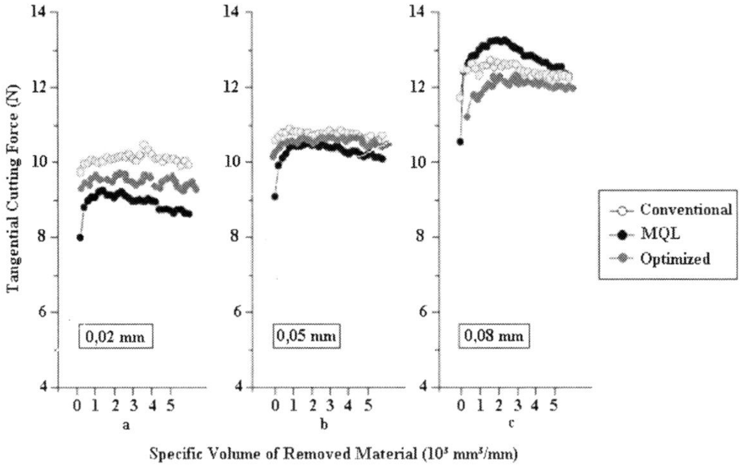

Figure 1: Tangential cutting force data – a) 0,02mm; b) 0,05mm; c) 0,08mm.

inefficient lubrication, the cutting zone cooling by the air flow is not able to minimize the excessive heat generation. In this case, the optimized method seems to be more efficient, because it allows better penetration in the cutting region and the removal of the excessive heat generated.

The optimized method showed good results due to its capability of cooling the cutting zone. This capability relies in the way the cutting fluid is applied. The high speed enables the fluid to penetrate the cutting zone, passing through the air barrier generated by the grinding wheel. The behaviour of this method throughout the increasing of the depth of cut is constant. The conventional method presents higher values, although its behaviour is constant too. These higher values are caused by the air barrier of the wheel, which the abundant flow of cutting fluid cannot surpass efficiently.

It is possible to conclude that increasing the depth of cut, the average values of tangential cutting force tend to increase as well. This is a consistent presumption, since an increase in the depth of cut produces a larger contact area between the grinding wheel and the workpiece, and although the number of grains in contact is also higher, the strength to remove more volume of material is increased, which consequently produces higher cutting forces.

Observing the cutting force behaviour, which varies along the experiment, it is possible to notice its inconstancy. It is due to the CBN super-abrasive grains friability. Along the process, the grains are worn out, losing their cutting edges, becoming flatter. This produces higher grinding forces, due to the decreasing of their cutting capacity and to the increasing of the drag process. Those worn out grains are released from the matrix, renewing the cutting edges. CBN grinding wheels have this capacity of self-sharpening, which strengthens their quality characteristic.

3.2 Surface roughness

It is important to analyze the surface roughness, because the surface finishing affects the fatigue resistance of the workpiece significantly, when under strength. It is still linked to lubrication and depends mainly on the size of the abrasive grain present in the grinding wheel, on the dressing conditions and material removal rate [8].

Thus, due to the importance of surface roughness on ground workpiece, the average surface roughness (R_a) was measured. In Figure 2, the values are given in micrometers, resulting from the arithmetic mean of all experiments with the same depth of cut.

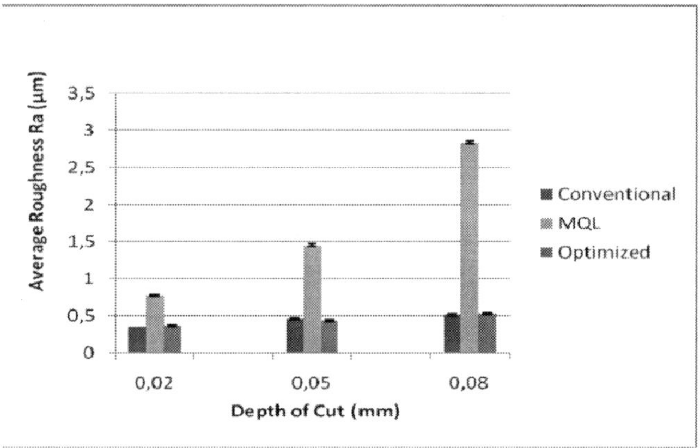

Figure 2: Comparison of roughness among the methods tested.

Figure 2 shows that all gathered in the conventional and optimized lubrication-cooling methods were below the ones obtained in the MQL method. At 0.08mm depth of cut, surface roughness data in MQL are almost six times higher than the other values.

According to Malkin [2], the acceptable range for average roughness in a grinding process is from 0.2 to 1.6μm. Therefore, all values obtained with the conventional and optimized methods are within the tolerance range, being quite satisfactory. The values are very similar, so it is possible to conclude that both methods remove chips away and guarantee the characteristic surface quality of the grinding process.

In the MQL, however, the data obtained at 0.08mm depth of cut are out of the considered range, which disqualifies the process concerning its surface quality. At the best condition, 0.02 mm, the values produced are higher than the ones produced by the other two methods. Nevertheless, MQL is a considerable alternative, even for finishing. For 0.05mm, MQL is in the limit of the considered range, which means the surface quality has not been kept. In this case and for 0.08mm depth of cut, the MQL technique can be disqualified.

The reasons of those variations can be explained. The conventional and optimized methods allow the removal of the chips in the cutting zone, improving the lubrication in the region and minimizing the friction between the tool and the workpiece. The MQL method, being inefficient in these tasks, makes the chips remain in the cutting zone, and the bigger the chip, more difficulty is to expel it, and worse surface quality is generated, which can be corroborated by the high values of surface roughness. The air flow is efficient in removing the chips produced with a 0.02mm depth of cut. However, its efficiency decreases with the increase of the depth of cut, fact confirmed by the 0.05mm value.

Although MQL requires lower cutting forces, the surface roughnesses are higher. The explanation to this fact consists of the grinding wheel topography. The CBN wheel lodges the thin chips produced in the MQL methods, resulting in lower forces and medium surface roughness. However, the increase in the depth of cut increases chip size, which makes more difficulty to settle the chips in the wheel porosity, which causes deplorable surface quality.

3.3 Grinding wheel wear and G ratio

Grinding wheel wear is an extremely important variable in grinding, since the more the wear, the shorter the wheel life cycle. Figure 3 presents the diametric wear of the grinding wheel at each depth of cut, for the three types of lubrication-cooling considered.

Figure 3 illustrates that the diametric wear of the grinding, wheel using the MQL technique, was lower at 0.02 and 0.05mm depths of cut, whereas at 0.08mm, the value obtained was much higher than with the other two methods. These behaviours can also be observed with the cutting force data, since the higher the strength on the grinding wheel, the higher forces of cut needed and, consequently, higher tool wear.

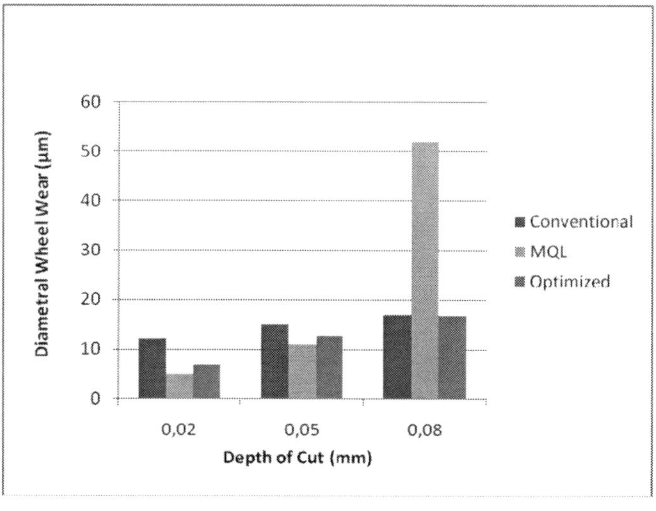

Figure 3: Grinding wheel wear comparison.

Undoubtedly, the more the lubrication-cooling fluid penetrates in the cutting zone, less wear is produced by friction better is the removal of the generated heat. The porosity of the grinding wheel should be considered, since it is responsible for the accommodation of the chips produced, improving the access of the fluid to the cutting zone, with consequent improvement in lubrication, minimizing thus the wear by friction.

The grinding wheel used is a CBN grinding wheel with vitrified bond, which is less susceptible to erosive wear by chips, presenting a greater retention force on the abrasive grain and, consequently, minimizing the diametric loss of the grinding wheel.

In MQL at low depths of cut, parts of the chips mixed with atomized oil lodge in the wheel pores, allowing a portion of the lubricant to reach the cutting zone more effectively, decreasing the wear. This also allows the air flow to remove part of the heat produced, improving the cooling. So, even with this lodging condition producing a higher surface roughness, it contributes to lower tool wear.

However, when the generated chips becomes bigger, as when increasing the depth of cut, the pores are not able to accommodate them, and the air flow is not efficient to remove them from the cutting zone. Then, the chips damage the cutting and increase the abrasive strength of the tool on the workpiece, decreasing the lubrication capacity in the cutting zone, prejudicing the surface and wearing the wheel more intensely.

The optimized method also presents satisfactory results, being much more efficient than the conventional. The high speed of fluid application makes the removal of chips and excessive heat easier in the cutting zone, than the abundant fluid flow. It justifies why the optimized method showed better results, guaranteeing longer wheel life.

The G Ratio is the ratio between the quantities of removed material and of worn material. This relation was obtained from the grinding wheel wear data.

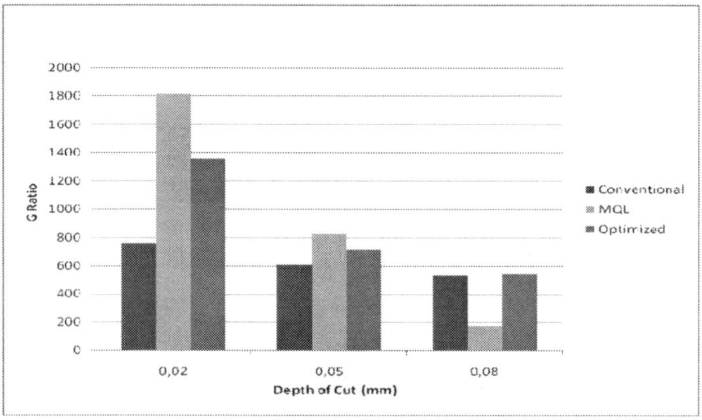

Figure 4: G Ratio comparison among the lubrication-cooling methods analyzed.

Figure 4 shows that the MQL is a good alternative concerning the application of lower depths of cut. The optimized method needs to be highlighted, since it shows that the fluid application in high speed is more efficient than the excessive fluid application in the conventional one. As the depth of cut increases, G ratio values decrease, due to the increase of tool wear. For medium values, the MQL and the optimized method were the best. In this condition, however, the MQL presents high values for roughness, which is harmful for the process.

For higher depths of cut, the MQL showed to be ineffective in its functions. In this condition, the performances of the conventional and optimized methods e were similar, because the high quantity of heat is difficult to be removed, even with the fluid being applied at high speed.

3.4 Microhardness

Microhardness in the ground workpieces was also analyzed. This kind of analysis is made in deeper regions of the surface. After analyzing the gathered data, the graph in Figure 5 was produced. It also shows the reference value adopted, which is the microhardness before grinding.

The microhardness values obtained in the workpieces show that in all machining conditions and using the three methods of lubrication-cooling there is a loss concerning the reference. This is associated with the capacity of heat dissipation and cooling speed of the specimens. Such factors are determinant to define the granular structure obtained.

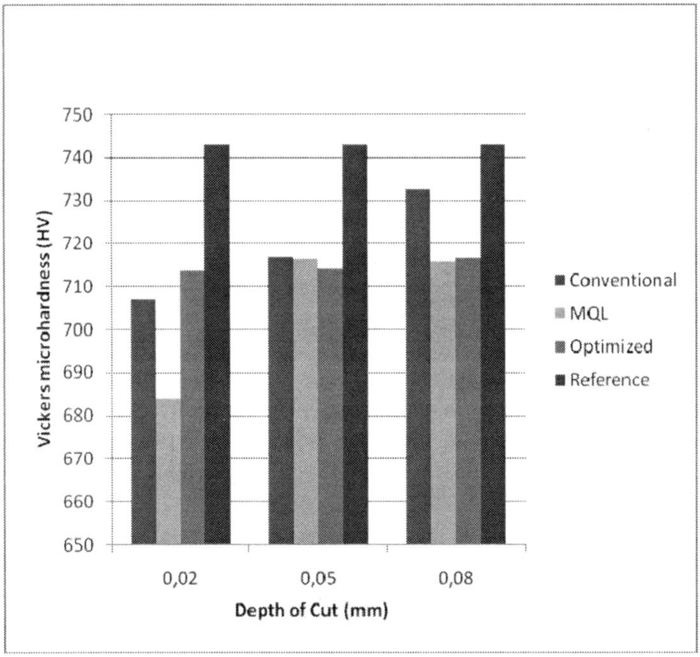

Figure 5: Vickers microhardness in the ground workpieces.

Analyzing firstly the data gathered with 0.02mm depth of cut, the loss obtained using the conventional method is smaller than the one using the MQL, but higher than the one with the optimized. This is consistent with the fact that MQL produces much more heat during cutting, while the excessive fluid and directed fluid jet of the conventional and optimized methods, respectively, remove more heat from the cutting zone. The ground workpieces under conventional lubrication-cooling lost around 5% of their surface micro-hardness, compared to the reference. The ones ground by the optimized lost around 4%, whereas by MQL, approximately 8%.

At 0.05mm depth of cut, the differences between the fluid application methods are very small, showing a very similar behaviour, which makes impossible to draw any conclusions about the microhardness tendency. On the other hand, there is a loss around 4%, compared to a non-ground workpiece.

At 0.08mm depth of cut, the values obtained by MQL are much lower than with conventional and optimized methods. This can be explained by the fact that MQL, in this condition, produces too much heat during cutting, consequently damaging the workpiece (with high surface roughness) and the tool (with enormous wear). In this grinding condition, microhardness loss by MQL was around 4% and approximately 1.5% by the conventional method. The optimized method was not very efficient to remove heat at high depths of cut, presenting behaviour similar to MQL and inferior to the conventional method.

Another interesting fact can be observed in figure 5. The microhardness loss in the most severe condition was smaller than the one obtained in the mildest condition. One of the reasons is that the latter took as many as three times longer than the most severe. Therefore, there was more exposure of the surface to constantly generated heat, although this was present in smaller values. Another explanation concerns to the performance of the CBN grinding wheel, developed for more severe work conditions, which allows better results under heavier machining conditions.

To sum up, it is possible to conclude that the heat generated during the grinding causes a reduction of the workpiece hardness, independently of the lubrication-cooling method utilized.

4 Conclusions

From data gathered with the experiments using three different methods of lubrication-cooling – conventional, optimized (with Webster nozzle) and MQL – it is possible to conclude that, for plane grinding of AISI 4340 tempered and quenched steel, varying the depth of cut while keeping other grinding parameters constant, that:

- The tangential cutting force presented the smallest results in the MQL for 0.02 and 0.05mm depth of cut, whereas for 0.08 mm in this method this variable suffers a high increase, being superior do the value obtained with the other methods.

- The surface roughnesses of ground workpieces by conventional and optimized methods were always smaller than the ones presented in MQL. Surface roughness of workpieces ground when applying MQL is still in the acceptable range; however, for medium and high depths of cut, is remains on the limit or exceeds the values considered as acceptable. Under these conditions, the MQL produced low quality surfaces.
- The grinding wheel wear was smaller in MQL than in the other two methods. The optimized method is to be highlighted for presenting good performance at low and medium depths, whereas the conventional method is to be highlighted for higher values.
- The workpieces microhardnesses showed that in the three methods there was a loss of hardness due to the heat generated during the cutting process.

The MQL technique in grinding showed itself viable only when the removal of material is not severe. This way, it is possible to use MQL without considerable surface quality loss and with production gain, due to the longer life cycle of the grinding wheel. However, surface quality is impaired, producing rougher surfaces. For more removal of material, the MQL shows it is not an efficient lubrication-cooling method, damaging the workpiece finishing, wearing too much the tool.

The optimized method is to be highlighted for low and medium depths of cut values. It produces excellent surface finishing and small wheel wear. It shows that the application of cutting fluid with high speed is efficient in removing chips and the excessive heat from the grinding zone. For higher depths of cut, the conventional method is better indicated, although the resulting parameters were not satisfactory.

The selection of the method depends on several parameters. As long as equipment is available, the MQL is a very effective method for low material removal. The optimized method with Webster nozzle is a good alternative to the conventional method, because it makes the fluid application more effective, allowing a longer life cycle for the grinding tool.

This analysis was made varying the depth of cut and keeping other grinding parameters constant. A proposal for further researches is to vary other parameters, in order to attend industries demand, and analyze whether the cooling methods are efficient in maintaining process quality.

References

[1] Chen, X.; Rowe, W.B.; Mills, B.; et al., 1998, "Analysis and simulation of the grinding process. Part IV: effects of wheel wear", International Journal of Machine Tools & Manufacture, Vol. 38, No. 1-2, pp. 41-49.

[2] Malkin, S., 1989, "Grinding Mechanisms e Grinding Temperatures and Thermal Damage", In: Malkin, S. Grinding Technology: theory and

applications of machining with abrasives, 1 ed., Chichester, Ellis Horwood Limited.

[3] Hassui, A.; Diniz, A. E., 2003, "Correlating surface roughness and vibration on plunge cylindrical grinding of steel", International Journal of Machine Tools & Manufacture, Vol. 43, pp. 855–862.

[4] Irani, R.A.; Bauer, R.J.; Warkentin, A., 2005, "A review of cutting fluid application in the grinding process", International Journal of Machine Tools & Manufacture, Vol. 45, pp.1696–1705.

[5] Kopac, J.; Krajnik, P., 2006, "High-performance grinding – A Review", Journal of Materials Processing Technology, Vol. 175, pp. 278-284.

[6] Tawakoli, T.; Westkämper, E.; Rabiey, M. et al., 2007, "Influence of the type of coolant lubricant in grinding with CBN tools", International Journal of Machine Tools & Manufacture, Vol. 47, pp.734-739.

[7] Ebbrell, S.; Woolley, N. H.; Tridimas, Y. D. et al., 2000, "The effects of cutting fluid application methods on the grinding process", International Journal of Machine Tools & Manufacture, School of Engineering, Liverpool, U.K., Vol. 40, pp. 209-223.

[8] Webster, J. A., 1999, "Optimizing coolant application systems for high productivity grinding", ABRASIVES Magazine, October/November, pp. 34-41.

Cavitation damage observations within scroll expander lubrication systems

I. Tzanakis[1], M. Hadfield[1], A. Georgoulas[2] & N. Kotsovinos[2]
[1]*Sustainable Design Research Centre, School of Design, Engineering and Computing, Bournemouth University, UK*
[2]*Department of Civil Engineering, Democritus University of Thrace, Greece*

Abstract

Observations of cavitation damage within a lubricated expander system are studied experimentally. Typical experimental analysis is used to observe cavitation erosion features such as SEM, light-microscopy and light-interferometer.

An experimental test-rig is used to study bubble characteristics within fluids to compare the theoretical analysis and practical in-service results from the expander system. Using an ultra-sonic methodology and high-speed camera techniques the bubbles are observed within the working fluids.

A 2D numerical simulation of the scroll was performed to explore the mechanism which generates scroll cavitation. It is found that the pressure is high enough to liquefy instantaneously part of the refrigerant close to the bottom boundary, creating conditions for the generation of cavitation bubbles within the liquefied refrigerant. This finding resolves the puzzle how the refrigerant which enters the scroll in gas phase produces cavitation.

Keywords: expander, scroll, refrigerant, lubricant, cavitation, erosion, bubbles.

1 Introduction

Scroll expander has been the main part of domestic appliances over the last few years (CHP systems, air-conditions, pumps etc). Many of these devices are facing cavitation wear problems because of their long operational period, their complicated geometry, their mechanical design and their high-speed working fluids. Interestingly, cavitation was identified by the author during the

www.witpress.com, ISSN 1743-3533 (on-line)
doi:10.2495/TD100221

investigation of a specific scroll device in a previous study [1]. A small numbers of papers, published the last few years, are dealing with the operation of the scroll expander in an organic rankine cycle (ORC) system. The papers are mainly focused on the performance and the efficiency of the scroll expander rather than looking the tribological aspects of scroll's critical components during its operational period [2–7].

In this study a scroll expander which was under operational conditions for a thousand hours, as a main part of a small domestic combined heat and power (CHP) organic rankine system (ORC), is experienced damage in its main components which without any doubt identified as cavitation damage. The fluid environment consists of the two scroll fluids: the high molecular organic fluid (refrigerant) which mainly is in a gas form and drives the scroll and the synthetic oil (lubricant) which protects the parts of the scroll from excessive wear. The puzzle is which of the two fluids and by which mechanism the cavitation damage is produced. At first glance the pressure and temperature conditions inside the scroll expander are such that the lubricant liquid will not has chance to change phase from liquid to gas (to produce bubbles) and the refrigerant which is entering scroll as gas at high temperature (around 160°C) is supposed to always remain in a gas state. Thus, the mean of generating cavitation by instantaneous transformation of phases from liquid to gas is not obvious. Not much of experimental data on the lubricant bubbles [8–10] and substantially fewer on the refrigerant bubbles [11, 12] relating to their velocity and impact attributes are available in literature.

The experimental study investigates the formation of the pits in their incubation period. The evaluation of the photographic results in correlation with the experimental results by the collapse impingement of the bubbles led to a more complete understanding of the cavitation erosion mechanisms. Using CFD methods the cavitation mechanism of the scroll expander was revealed. A 2D analysis of the scroll was performed. The aim was to investigate the behaviour of the refrigerant jet stream as it develops inside the first expansion chamber of the scroll. On the basis of numerical simulation results, it is advocated that the cavitation mechanism is related to the initial instantaneous high pressure at the lower scroll boundary induced by the refrigerant inflow gas jet. Instantaneously part of the refrigerant gas phase is transformed to liquid phase due to high pressure, which subsequently due to turbulent environment and high pressure fluctuations produces cavitation bubbles within the refrigerant in liquid phase.

2 Experimental methodology

Cavitation erosion tests were utilized in the present study to analyse and evaluate the performance of the cavitation bubbles generated within the two scroll working fluids. The experiments were performed according to the ASTM G32-03 standard method. The ASTM method proposes a standard probe of 15.9 mm diameter to be generally used for cavitation tests. However, in this study a probe of 5 mm was deployed to produce cavitation bubbles.

The steel plate of the scroll is machined into appropriate samples shape by a cutting machine. Then the steel samples were implemented on a Bakelite base. All the test samples were polished in a few microns accuracy and went through a surface analysis to ensure that they were free from any surface defects prior to testing. The chemical composition of the steel identified with the use of an optical emission spectrometer is shown in table 1.

Table 1: Chemical analysis of the scroll's steel plate.

Steel Plate	%C	%Mn	%Si	%S	%P	%Cu	%Ni	%Cr	%Fe
Wt%	0.96	0.53	0.27	0.003	0.021	0.01	0.02	0.18	REM

The experimental apparatus is described in figure 1. Experiments were carried out by using an ultrasonic transducer at a frequency of 20 kHz. Ultrasonic horn has a plane surface oscillating in a simple harmonic motion. Peak to peak amplitude of the horn tip was adjusted to be at 50 μm. The sample was mounted on the bottom of a small transparent tank with a maximum capacity of 500ml. The ultrasonic horn was submerged into the liquid environment of the tank in a distance of 0.5 mm from the chromium steel surface. Furthermore a high speed camera (Phantom v7.3) was installed at one of the sides of the tank. At the opposite side of the tank a high intensity fibre optic light source was implemented. Pictures were taken in random time steps during the cavitation process.

The liquids used for the tests were a synthetic lubricant and a high molecular organic refrigerant. All the liquids were tested in a room temperature of around 23±2°C. The physical properties of the fluids used for the experiments are listed

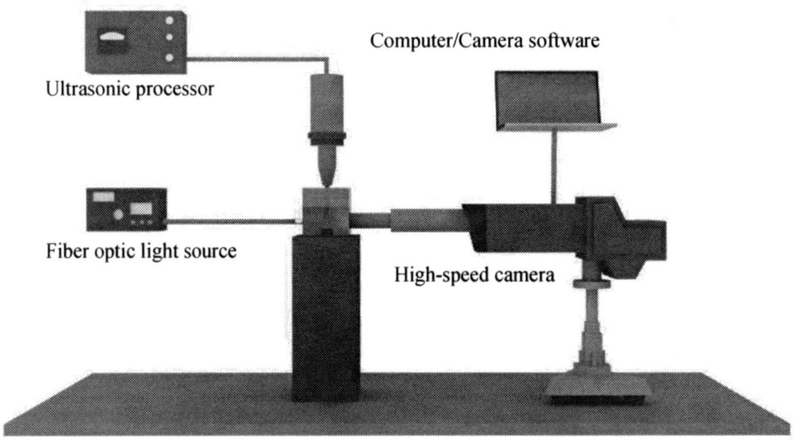

Figure 1: Schematic of experimental set up.

Table 2: Physical properties of the experimental liquids at 25°C/1bar.

Fluid	Refrigerant	Lubricant
Ultrasonic Velocity (m/sec)	990	1700
Liquid Density (kg/ m^3)	625	995
Viscosity (mPa·s)	0.225	30
Surface Tension (mN/m)	32-35	49-54
Vapour Pressure (Pa) at 20°C	53300	1300

in table 2. The viscosity and surface tension for all the three fluids was measured using appropriate experimental devices.

Selected testing time was ten minutes, in order to prevent the temperature of the tested fluids from rising more than two to three degrees during the cavitation process. Furthermore, the amount of the incubation pits formed on the surface of steel samples was sufficient enough for analysis and interpretation. The temperature was controlled every minute with a laser digital thermometer.

3 Surface investigation

After the end of the process, the scroll expander of the CHP unit was disassembled and its high carbon steel plates were isolated and sent for surface investigation. This procedure was followed in order to analyse thoroughly the surface of the steel plate. Three different types of microscopes, an optical, an electronic (SEM) and an interferometer (ZYGO) microscope were deployed.

Interestingly, during the surface analysis of the scroll's steel plate, cavitation erosion was easily identified under the optical microscope. Many cavitation pit clusters were spotted across the high pressure regions of the steel plate, in the beginning of the expansion process. The cavities were mainly accumulated along the edges of the steel plate, creating a continuous layer of cavities of about 2mm wide in both edges (figure 2). Surprisingly, on the low pressure regions of the steel plate, at the end of the expansion cycle, no trace of cavities was found.

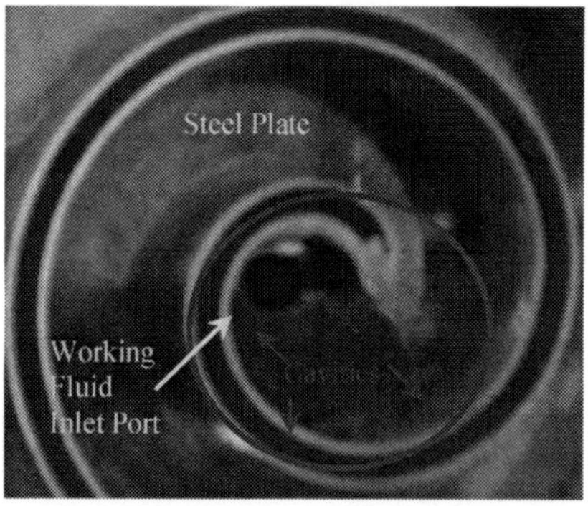

Figure 2: The high pressure area of the scroll expander. cavities formed across the edges (red arrows).

The average width of the steel plate is around 12.5mm. However, in the areas where cavitation was revealed the width is even less, due to the geometrical profile of the scroll. Thus, the 2mm width range of cavitation across the edges of the steel plate can be considered as a severe surface damage. The existence of cavities can rapidly increase the cavitation process creating deeper holes hence increasing the leakage and reducing the efficiency of the scroll.

In figure 3 a typical cavity found on the steel plate is analysed. The average depth of the individual cavities was estimated around 6-8 microns. In some cases the depth of the cavities was up to 10 microns. In addition, craters were formed due to a consistent and rigorous impact from the bubbles implosion inside a single cavity. Their enlargement into craters is a time and history dependent process.

a) b)

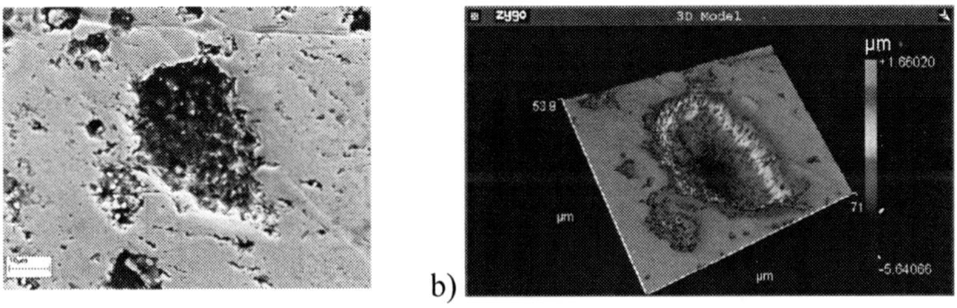

Figure 3: Cavities profile in detailed a) SEM microscope b) ZYGO interferometer.

4 Test results

The results produced were distinguished in two categories: the numerical and the experimental. The rational behind this was a coherent interpretation of the erosion mechanisms found on the steel plate surface. Initially, the numerical results produced with computational fluid mechanics (CFD) methods, offered by the commercial CFD software FLUENT, explain the generation of cavitation bubbles inside the scroll expander. The experimental results reveal the performance of the working fluids and their impact mechanisms on the actual steel plate of the scroll.

4.1 Numerical results

With the use of CFD methods, a finite volume numerical approach was applied in order to reveal the cavitation mechanism of the scroll expander. A 2D analysis of the scroll was performed. The aim was to investigate the behaviour of the jet stream as it develops inside the first expansion chamber of the scroll. It was believed that the cavitation mechanism was derived by the low pressure at the lower scroll boundary induced by the refrigerant. Thus, a drop of pressure by the entrainment was thought to be sufficient enough to make the lubricant volatile. Interestingly, only a small pressure drop was observed in the entrainment points while a tremendous increment of pressure was spotted in the scroll.

The numerical model that was used in the simulations solves the Reynolds Averaged Navier Stokes equations (RANS). Closure of the turbulent stress terms is achieved with the application of the standard k-ε turbulence model. In more detail the equations solved are the continuity equation, the momentum equation, the energy equation and the transport equations for the turbulent kinetic energy (k) and its dissipation rate (ε), respectively. At the solid boundaries of the geometry a wall boundary was applied, with enhanced wall treatment for the prediction of the turbulent quantities at the near-wall regions. A velocity-inlet boundary condition was used for the inlets and a pressure-outlet boundary condition for the outlets.

Two critical areas were chosen for the analysis. The first area is the moving wall (right side-outflow) at its minimum distance from the inlet port (inflow) in the very first stage of the expansion process. The second area is at the maximum operating distance of the moving wall (outflow) from the inlet port. In the first area a total number of 20493 quadrilateral cells were used, while the second area consists of a total number of 22050 quadrilateral cells. In both cases, grid clustering was used near the boundaries (figure 4).

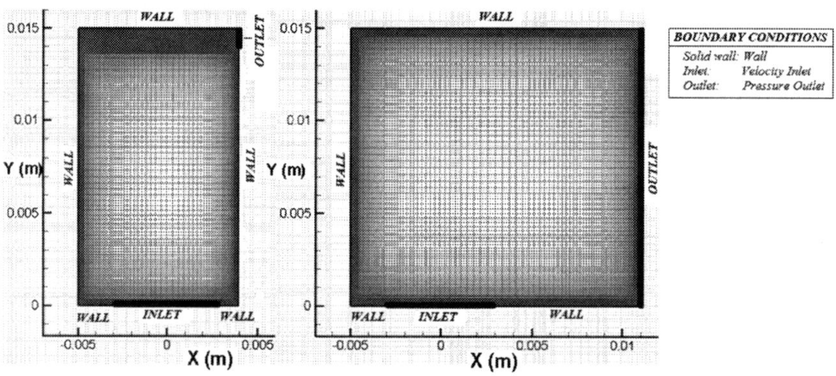

Figure 4: Geometry, computational mesh and boundary conditions of the first expansion pocket of the scroll expander.

At the early stage of the scroll operational mode there are no gaps created. Therefore, refrigerant is prevented from passing to the next pressurised chamber as it is introduced to the scroll. The refrigerant thrust pushes the scroll wall creating a gap. Thus, in the first case of the numerical analysis, a small gap was used for the equation of continuity to be solved. This was the worst scenario for generating high pressure fluctuations inside the scroll. In the second case, where the optimum scenario for avoiding cavitation due to pressure fluctuations is used, the gap is assumed to be across the whole side of the scroll wall, when the moving wall (outflow) is in its maximum possible operating distance from the inlet port. In both scenarios the shape and the operational conditions of the scroll expander were adjusted to the real world case, for representative results. Each pressure cycle according the operational mechanism of the scroll lasts around 17 msec.

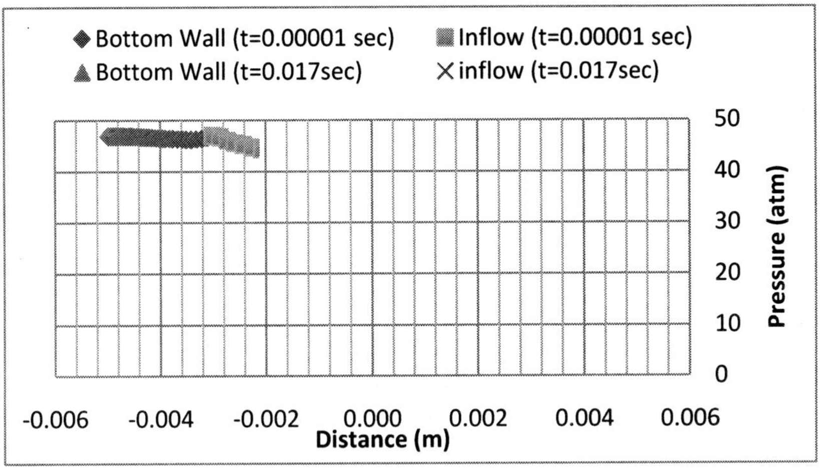

Figure 5: Pressure distribution along bottom boundary (initial position of moving part).

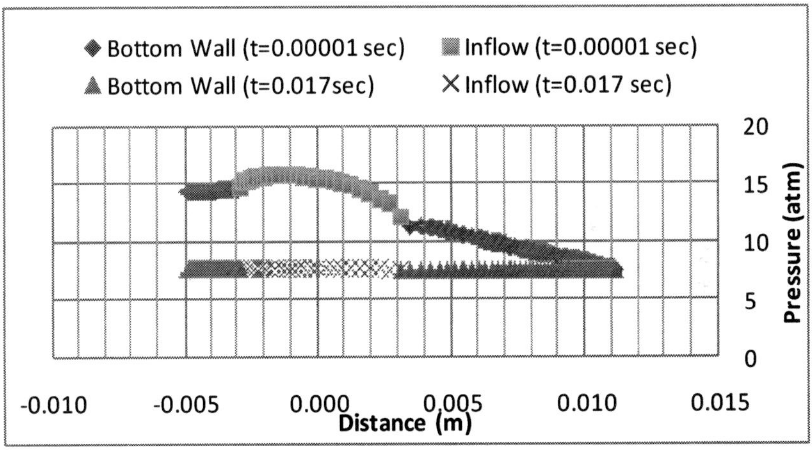

Figure 6: Pressure distribution along bottom boundary (final position of moving part).

The velocity of the inflow gas jet refrigerant is 31 m/sec. Subsequently in figures 5–6 the instantaneous pressure distribution for t=1 μsec and t=17 msec is presented, on the bottom plate where cavitation was found. From both figures 5–6, for t=1 μsec, when the jet stream from the refrigerant is developed, there is a steep increment of the ambient pressure. In figure 5 the refrigerant pressure reaches for t=1μsec an extremely high level of around 45 bars across the bottom surface and for t>1 μsec the pressure decreases and is stabilized at around 17 bars until the end of every pressure cycle.

The same mechanism applies when the moving wall is at the maximum operating distance from the inlet port. As figure 6 shows, the ambient pressure across the bottom wall increases to 16 bars and then drops down to 8 bars. In both cases the pressure is high enough and the transformation of the lubricant from the liquid to gas phase is not possible. Therefore it seems that the lubricant is not responsible for the cavitation. However, in both cases the pressure is high enough to liquefy instantaneously part of the refrigerant close to the bottom boundary, creating conditions for the generation of cavitation bubble within the liquefied refrigerant. This finding resolves the puzzle how the cavitation erosion mechanism on the steel plate is produced.

4.2 Experimental results

The photographic results from the high speed camera revealed some interesting images regarding the performance of the lubricant and the refrigerant cavitation bubbles.

In the case of the lubricant, the bubbles were generated in low pressure areas and then collapsed in regions where the pressure was higher, creating severe damage on the surface of the material. Interestingly, a lubrication layer is generated during the cavitation process (figure 7). The lubrication layer can affect the implosion phenomenon of the bubbles. It can protect the material surface by acting as a cushion, reducing the impact from the bubbles implosion. Thus, despite the fact that the impact pressure from the lubricant bubbles is higher than in the refrigerant, the actual damage can be considerably less.

When refrigerant was used, bubbles had a remarkable behaviour while approaching the surface of the steel plate in figure 8. They sustained a continuous oscillatory movement without imploding. A ceaselessly wobbling effect was also noted. Eventually the refrigerant bubbles obtained an amoeboid shape. Most of the refrigerant bubbles wobble instead of collapsing. Eventually no significant material loss or severe surface damage was caused. The surface tension of the refrigerant could possibly play an important role in this behaviour.

At the very early stages of the cavitation process, the impact energy released during cavity collapse on the surface is absorbed without any material deformation. Then, plastic deformation occurs which propagates into the grain boundary of the material displacing the grains' structure. Subsequently, within the ten minutes of the experimental process, the boundary grains are fractured, material loss is observed and cavitation pits are formed.

In figure 9 a typical destructive cavitation pit, produced in a lubricant and a refrigerant environment, is shown. The SEM image clearly illustrates the nature of the pit and its shape. The depth is close to the average typical depth produced by the impact of the lubricant and refrigerant bubbles respectively. The determinant between the two different types of pits is their depth level.

This is clearly depicted in figure 10. The pits produced on the surface of the steel sample within the lubricant environment are on average 5 times deeper than the ones found on the refrigerant environment. In addition, the average depth of the larger pits produced by lubricant bubbles was found to be three times higher

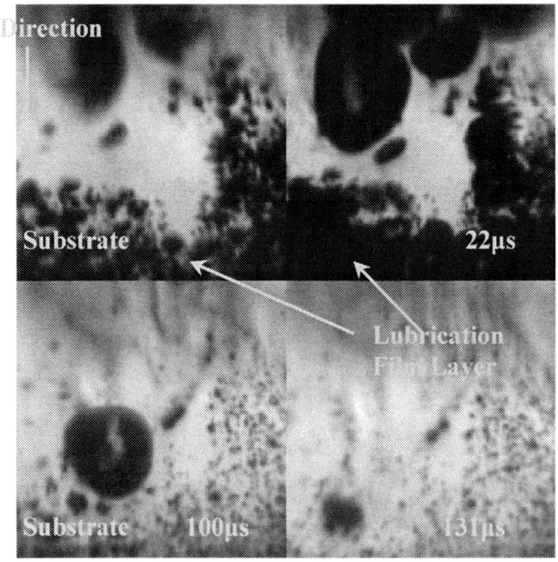

Figure 7: Collapse of an acoustic cavitation lubricant bubble near the solid boundary. A lubrication layer is formed. The frame size is 0.5×0.5mm [12].

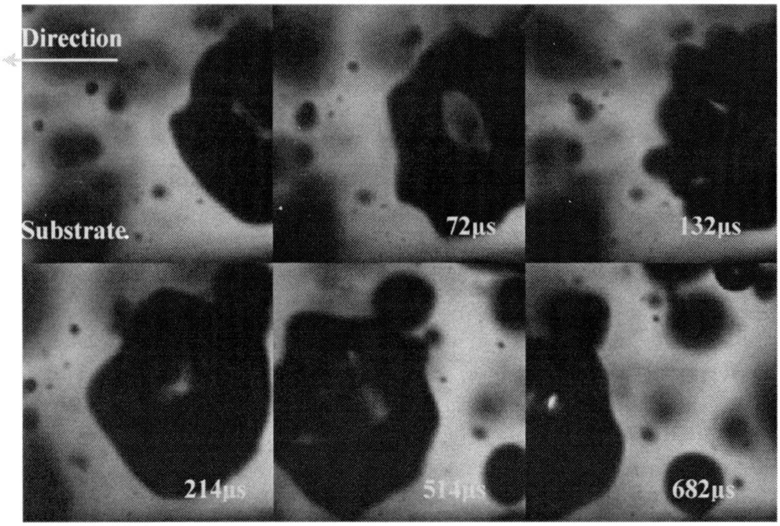

Figure 8: The ceaselessly wobbling movement of an acoustic cavitation refrigerant bubble near the solid boundary. The frame size is 0.5×0.5mm [12].

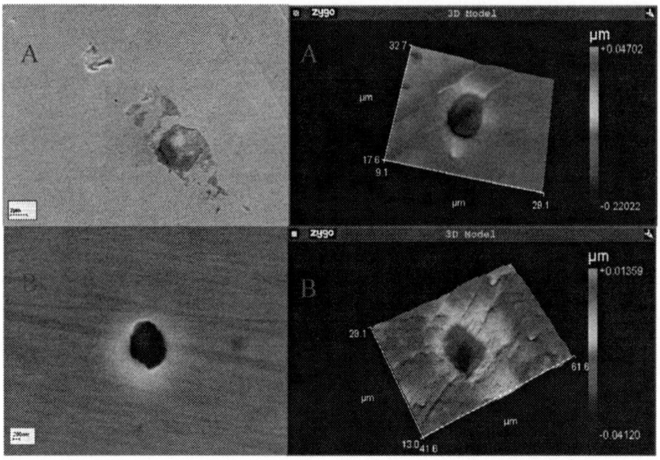

Figure 9: A typical erosion pit a) lubricant b) refrigerant.

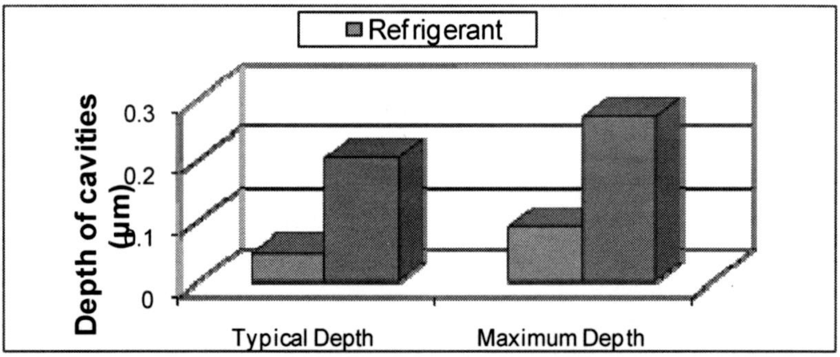

Figure 10: Depth of cavitation pits in different environments.

in comparison to the cavities generated by the implosion of the refrigerant bubbles.

5 Discussion

After the thorough surface investigation and the evaluation of the experimental and the numerical results, the cavitation mechanism responsible for damaging the surface of the scroll's steel plate is revealed. The initial consideration that cavitation derives from the lubricant, because its liquid phase was easier to generate bubbles, seems to weakening after the results produced from the computational fluid dynamic analysis. Interestingly, the refrigerant bubbles seem to be the main reason for the generation of cavities during the operational period of the scroll expander. The refrigerant bubbles are imploded near the boundary surfaces damaging the material.

The refrigerant working fluid while entering the scroll and striking to the upper boundary produces a tremendous impact, resulting to a steep increase of the ambient pressure. This increment of pressure is sufficient enough to change its phase from gas to liquid. Moreover this ambient high pressure lasts only for 20 to 30 μsec. However, this time, having observed the generation of bubbles through the high speed camera, is adequate enough to produce cavities inside the bulk environment of the scroll.

Finally, the refrigerant molecules as they change phase from gas to liquid can accelerate the miscibility process with the lubricant inside the scroll. The vapour point of the lubricant can be significantly reduced according to the miscibility degree. At high miscibility degree, the lubricant becomes more volatile which in turn reduces the performance of the lubricant film thickness which results in increasing abrasion. Future work in this area will involve a systematic study of the refrigerant lubricant combination in order to define the physical properties under various temperature and pressure conditions.

6 Conclusions

It is reasonable to conclude from the results of this study that the responsible cavitation mechanism for the erosion of the scroll's steel plate was determined. Refrigerant is the fluid mean which generates cavitation bubbles due to elevated pressure fluctuations inside the high pressure gas pockets of the scroll expander as the analytical results revealed.

This finding resolves the puzzle how the refrigerant which enters the scroll in gas phase produces cavitation. This is really an exceptional situation. Usually we have cavitation in an environment where the fluid in motion is in liquid phase where at low pressure is transformed to gas producing cavitation bubbles. In our case, we have a situation that the fluid in motion is in gas phase, which under very high pressure conditions is locally liquefied, producing cavitation bubbles within the high pressure region. We are not familiar for a similar finding in the literature.

The experimental results from the ultrasonic transducer showed that the jet impact produced by the implosion of the refrigerant bubbles is less destructive compare to the lubricant bubbles. The depth of the pits is more significant in the lubricant environment compared to the refrigerant. Overall, refrigerant bubbles have a low impact pressure. However, over a prolonged period of time (1000 hours) the surface investigation results on the steel plate revealed the critical damage produced by the refrigerant bubbles.

References

[1] I. Tzanakis, M. Hadfield, Z. Khan. 2009. *Durability of domestic scroll compressor systems.* WIT Press 62 (229-240).

[2] M. Kane, D. Larrain, D. Favrat, Y. Allani. 2003. *Small hybrid solar power systems*. Energy 28 (1427-1443)

[3] T. Yanagisawa, M. Fukuta, Y. Ogi, T. Hikichi. 2001. *Performance of an oil-free scroll-type air expander*. Proceedings of the IMechE Conference on Compressors and their Systems. (167-174).
[4] B. Aoun, D. Clodic. 2008. *Theoretical and experimental study of an oil-free scroll type vapour expander*. Proceedings of the Compressor Engineering Conference Purdue. (Paper 1188).
[5] S. Quoilin, V. Lemort, J. Lebrun. 2009. *Experimental study and modelling of an Organike Rankine Cycle using scroll expander*. Applied Energy.
[6] H.J. Kim, J.M. Ahn, S.O. Cho, K.R. Cho. 2008. *Numerical simulation on scroll expander-compressor unit for CO2 trans-critical cycles*. Applied Thermal Engineering 28 (1654-1661).
[7] V. Lemort, S. Quoilin, C. Cuevas, J. Lebrun. 2009. *Testing and modelling a scroll expander integrated into an Organic Rankine Cycle*. Applied Thermal Engineering 29 (3094-3102).
[8] E.A. Brujan, P.R Williams. 2006. *Cavitation phenomena in non-Newtonian liquids*. Chemical Engineering Research and Design 84 (293-299).
[9] P.R Williams, P.M. Williams, S.W.J. Brown. 1998. *A study of liquid jets formed by bubble collapse under shock waves in elastic and Newtonian liquids*. Journal of Non-Newtonian Fluid Mechanics. 76 (307-325).
[10] Y. Meged, C.H. Venner, W.E. ten Napel. 1995. *Classification of lubricants according to cavitation criteria*. Wear 186 (443-453).
[11] B. Schneider, A. Kosar, Y. Peles. 2007. *Hydrodynamic cavitation and boiling in refrigerant (R-123) flow inside microchannels*. International Journal of Heat and Mass Transfer 50 (2838-2854).
[12] I. Tzanakis, M. Hadfield. 2010. *Observations of acoustically generated cavitation bubbles within typical fluids applied to a scroll expander lubrication system*. Journal of Hydrodynamics.

Phosphonium ionic liquids as lubricants for aluminium-steel

A. E. Somers[1], P. C. Howlett[1,2], J. Sun[1,2], D. R. MacFarlane[2] & M. Forsyth[1]
[1]*Department of Materials Engineering, Monash University, Australia*
[2]*School of Chemistry, Monash University, Australia*

Abstract

The performance of a series of novel room temperature ionic liquids (ILs) based on the trihexyl(tetradecyl)phosphoniumcation (P_{66614}^{+}) and a number of novel anions have been studied in pin-on-disk tests using a 100Cr6 steel ball on AA2024 aluminium disks. The anions coupled to the (P_{66614}^{+}) cation include diphenyl phosphate (DPP-), dibutyl phosphate (DBP-), bis (2,4,4-trimethyl pentyl) phosphinate (M_3PPh^-) and bis(2-ethyl hexyl) phosphate (BEH-). More traditional anions such as bis(trifluoromethanesulfonyl) amide (NTf_2^-) and bromide (Br-) were also investigated.

Experiments were conducted at various loads to assess the IL film forming abilities. The results suggest that the structure of the anion is important in forming a surface film that reduces the friction and wear of the aluminium disk. At 30N five of the six ILs tested showed a 30-90% reduction in wear, as determined from wear scar depth measurements, compared to fully formulated diesel oil. The IL lubricant with a diphenyl phosphate anion achieved the lowest wear coefficient, showing a better performance than a typical fluorine-containing IL anion, NTf_2. To further investigate wear mechanisms and surface interactions the wear scars were analysed using a scanning electron microscope (SEM) and energy dispersive spectroscopy (EDS).

Keywords: ionic liquids, phosphonium, phosphorus anions, phenyl rings, lubricants, tribology, wear test, SEM, EDS.

WIT Transactions on Engineering Sciences, Vol 66, © 2010 WIT Press
www.witpress.com, ISSN 1743-3533 (on-line)
doi:10.2495/TD100231

1 Introduction

The use of aluminium alloys in technical applications is increasing due to their high strength to weight ratio, corrosion resistance and high thermal conductivity [1-3]. Unfortunately, aluminium alloys perform poorly in moving contact as adhesive wear and seizing is likely to occur [4]. For many current applications this means aluminium cannot be used unless it is coated with a wear protecting material, increasing the cost and complexity of production. It would be advantageous for certain applications if suitable lubricants were developed such that steel could move in contact with aluminium without causing excessive wear.

According to lubricant theory, at low loads oil additives with long polar molecules that are adsorbed onto the surface of moving parts work well to reduce friction and wear [5]. Researchers have shown that as the alkyl chain length of an imidazolium ionic liquid is increased the wear and friction between steel and aluminium are reduced [6, 7]. At increased loads extreme pressure additives in oils, such as zinc-dialkyl-dithiophosphate (ZDDP) for steels, are thought to react and breakdown in the presence of the fresh metal surfaces as a result of the increased temperature and pressure. These reactions form low friction layers, called extreme pressure (EP) layers, which reduce wear and friction. Unfortunately ZDDP has been shown to be incompatible with aluminium alloys as the film it forms can break up to form abrasive particles and a viable alternative is yet to be found [6, 8]. It has been suggested that the presence of elements such as phosphorus, fluorine, sulphur, boron, oxygen and nitrogen may be important for the formation of extreme pressure layers on aluminium alloy surfaces [2, 4, 9, 10]. However, depending on the IL structure, they can also lead to severe tribocorrosion [9].

Most common salts have very high melting points, for instance sodium chloride melts at 801°C. This high melting point is due to the strong electrostatic forces between the ions, each of which has a strong localised charge. Due to the structure of their molecules the charges on the ions of an IL are much more diffuse and so the electrostatic forces between the anion and cation are reduced such that they exhibit low melting points. Initial interest in ILs focused on their use as electrolytes and as non-volatile solvents for synthesis and chemical reactions but researchers are now investigating other suitable applications, such as lubricants and anti-corrosion coatings and inhibitors [11, 12]. The specific properties of ILs that make them suitable as prospective lubricants are their low volatility, so that they can be used in reduced pressure applications and their non-flammability and thermal stability, meaning they will be able to safely withstand increased temperatures involved when there is high friction. Since ILs have good solubility with organic compounds they can be used with current oils and additives and there is no need for detergents, defoamers or anti-oxidants to improve their compatibility [13, 14]. Another of the main attractions of ILs is the variety of molecules that can be used; one estimate is that there is something in the order of a million combinations available, each with its own unique properties [15]. This means that ILs may be highly tuneable for a particular application, particularly with respect to surface interactions. ILs and their production methods are also considered to be more environmentally friendly when compared to current alternative solvents and electrolytes and it is hoped that this will be the case for many new applications [14].

Ye et al. [16] were the first to assess the performance of an IL as a lubricant for various systems and interest has been increasing ever since [17]. Current research on ILs as lubricants is aimed at identifying those that will improve the wear performance of various systems, including aluminium in contact with steel and identifying the tribochemical reactions at the surface. For the steel-aluminium system, researchers have investigated ILs with various cations, such as imidazoliums [13, 14, 16, 18, 19], pyridiniums [13], ammoniums [14] and phosphoniums [4], and various anions, such as BF_4, PF_6, triflate, tosylate and NTf_2.It has been suggested that alkyl chain length, polarity and the elements present are important factors affecting the anti-wear performance of an IL [2, 3, 7, 9, 15].

Of the ILs tested thus far for the aluminium/steel system the vast majority have included fluorine-containing anions, the only exception being the tosylate anion ($CH_3C_6H_4SO_3$), which gave mixed results [7]. In this study the wear performance in a pin-on-disk test of ILs with a trihexyl(tetradecyl) phosphoniumcation (P_{66614}^+) and six different anions will be investigated at different wear regimes. Of the six anions to be tested, four have various novel structures as lubricants, such as different alkyl chain lengths, branched alkyl chains and phenyl rings and all contain phosphorus but no fluorine. As a comparison ILs with a fluorine containing NTf_2anion and a bromide anion were also tested. The system studied uses an ISO100Cr6 steel ball on an AA2024 aluminium disk in a rotating pin-on-disk test.

2 Experimental

Scheme 1 shows the structures of the ILs tested. The preparation of the ILs with the DPP, DBP, M_3PPh and BEH anions is detailed elsewhere [20-22], while Cytec Canada Inc. supplied the ILs with $NTf_2$and Bromide anions. The diesel oil used as a standard lubricant was a fully formulated 15W-50 multigrade oil meeting the API CH-4/SL standard.

The wear tests were conducted on a Nanovearotating pin-on-disk tester using 6mm ISO100Cr6 steel balls on AA2024 aluminium disks. The disks were lubricated with 0.1mL of the IL to be tested. Tests were conducted at loads of 10, 20 and 30N for a distance of 2500m, with a wear track diameter of 20mm and a speed of 0.2m/s. The coefficient-of-friction was recorded throughout the tests. On completion of the wear tests the wear depth was measured using a Dektak 150 stylus profilometer and these values were used to calculate the wear coefficient for each IL. Optical micrographs of the 100Cr6 steel balls were acquired using a Nikon Eclipse ME600. For more detailed analysis of the worn aluminium disk surface SEM and EDS results were obtained on a JEOL JSM-840A scanning electron microscope.

Table 1 shows values for the viscosity and conductivity of the ILs at 40°C.

Cation:

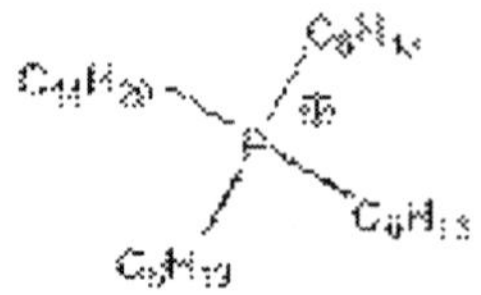

Trihexyl(tetradecyl)phosphonium (P_{66614}^{+})

Anions:

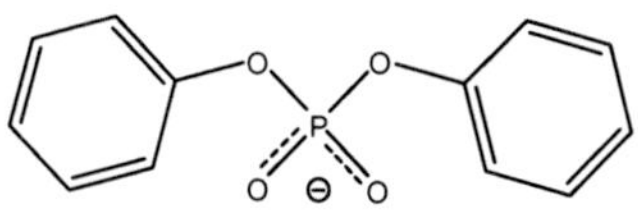

DiphenylPhosphate(DPP$^-$)

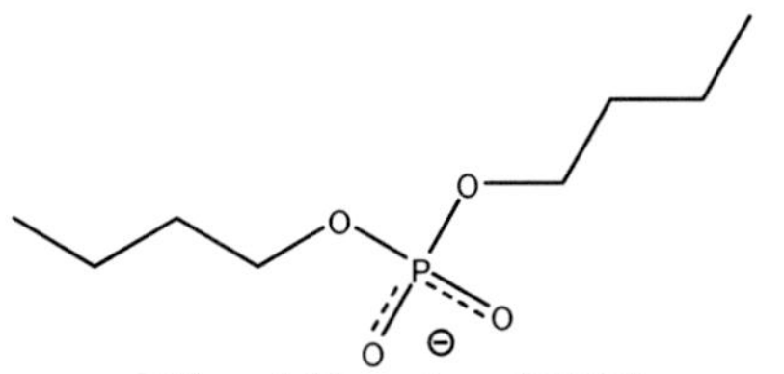

Dibutyl Phosphate(DBP$^-$)

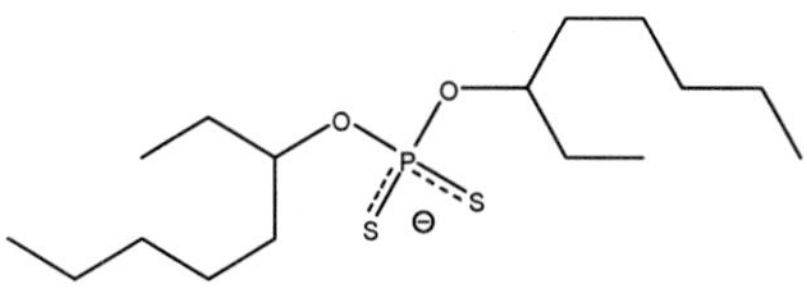

Bis-(2-Ethylhexyl)Phosphate Bis
2,4,4-Trimethylpentylphosphinate
(BEH$^-$)

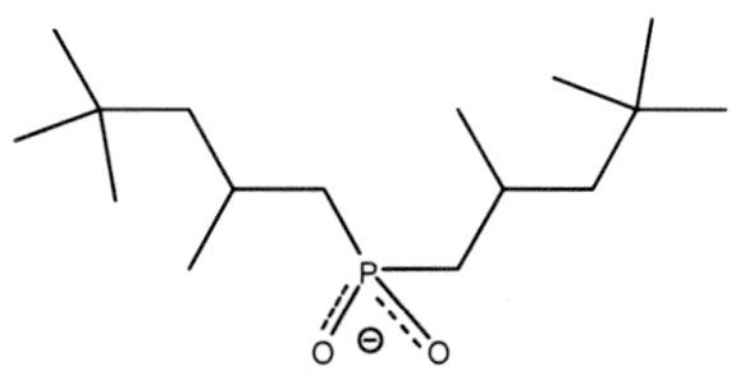

(M_3PPh^-)

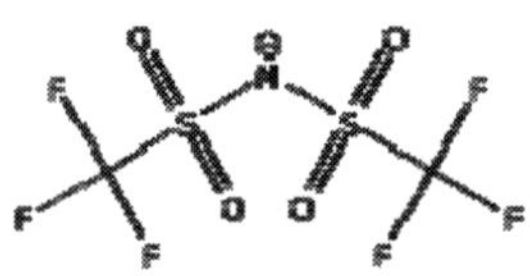

Bis(trifluoromethanesulfonyl) amide
(NTf_2^-)

Br^-

Bromide

Scheme 1: Ionic liquid structures.

Table 1: Ionic liquid viscosity and conductivity.

Ionic Liquid	Viscosity at 40°C (mPa.s)	Conductivity ($S.cm^{-1}x10^{-5}$)
P_{66614} DBP	130	2.0
P_{66614} M_3PPh	120	1.8
P_{66614} BEH	260	1.2
P_{66614} DPP	210	3.2
$P_{66614}NTf_2$	140	19.7
P_{66614} Br	610	2.3
15W50 Oil	340	0.000086

3 Results and discussion

3.1 Wear tests

Figure 1 shows the coefficient of friction (Fig. 1a) and wear coefficient (Fig. 1b) as a function of the load for the six ionic liquids and one fully formulated diesel

(a)

(b)

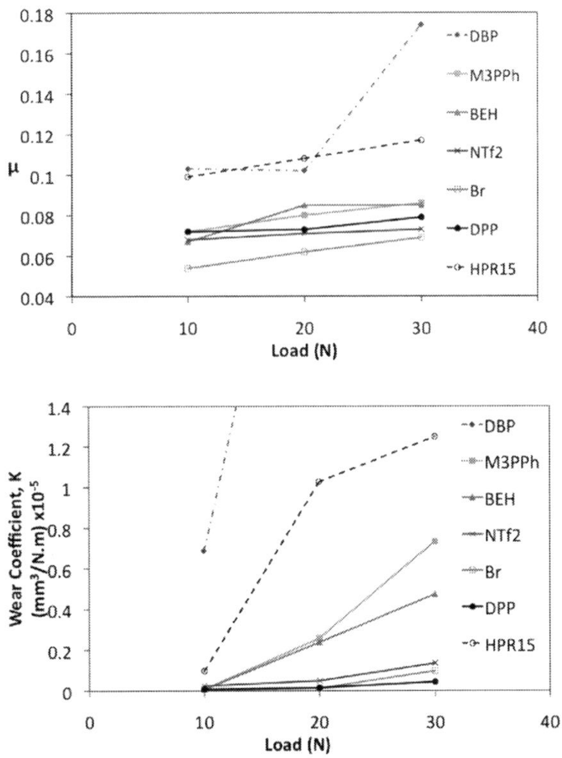

Figure 1: (a) Friction and (b) wear coefficient as a function of load.

engine oil tested. In both figures it can be seen that the P_{66614}DBP and the 15W50 engine oil lubricated samples have the highest friction and wear coefficients. In fact, the samples lubricated with P_{66614}DBP had such high wear coefficients at loads of 20N ($3.19x10^{-5}$) and 30N ($6.71x10^{-5}$) that they were not shown on Figure 1b so that the other results could be seen more clearly.

For the other ILs tested the friction coefficients results are much closer, with P_{66614}Br showing the lowest friction throughout, and P_{66614}DPP and $P_{66614}NTf_2$ consistently lower than P_{66614}BEH and $P_{66614}M_3$PPh.

The differences in the wear coefficient (Fig. 1b) are much more marked than the friction coefficients. As mentioned above, the P_{66614}DBP lubricated aluminium disk showed severe wear and the samples lubricated with the engine oil had significantly higher wear coefficients than the other samples. Of the remaining ILs the P_{66614}BEH and $P_{66614}M_3$PPh showed much higher wear than the $P_{66614}NTf_2$, P_{66614}Br and P_{66614}DPP. It can be seen that the aluminium disks lubricated with these latter three ILs resulted in very low wear coefficients. At 30N the wear coefficients for $P_{66614}NTf_2$, P_{66614}Br and P_{66614}DPP, respectively, are: $0.14x10^{-5}$, $0.10x10^{-5}$ and $0.04x10^{-5}$ mm^3/N.m.

3.2 Scanning electron microscopy (SEM)

Figure 2 shows SEM images of the wear scars on the aluminium disk and optical images of the scar on the steel balls for all the systems tested.

The pictures and spectra are shown in order of highest to lowest wear coefficient (see Figure 1b). For the sample lubricated with P_{66614}DBP the wear scars on the aluminium disk and the steel ball show much evidence of adhesive wear as there appears to be aluminium stuck to the surface of the steel ball. Moving through the samples the amount of adhesion can be seen to be reducing and on the P_{66614}Br and P_{66614}DPP lubricated samples there appears to be no adhesion occurring. On close inspection of the steel ball sample for the P_{66614}Br sample there are signs of what may be pitting corrosion above and below the wear scar, indicating that the steel ball may be corroding in the presence of the IL. This may explain why this sample showed the lowest coefficient of friction (see Figure 1a), as the corrosion product from the steel may have been forming a low friction film between the steel ball and the aluminium disk.

3.3 Energy dispersive X-ray spectroscopy (EDS)

Figure 3 shows EDS spectra taken of the IL lubricated aluminium wear scars tested at 30N. For the disks lubricated with P_{66614}DBP and P_{66614}Br the spectra show aluminium, oxygen, copper and magnesium from the alloy present, but no sign of any IL components. This does not mean there is not any IL present, as the films formed by ILs may be too thin for EDS to detect. For $P_{66614}M_3$PPh and P_{66614}BEH a small amount of phosphorus was detected, suggesting that an adsorbed layer of IL was present. Since these lubricants showed a moderately high wear coefficient, there may be some tribocorrosion occurring, where the IL reacts with the surface to such an extent that it is removed as corrosion product, thus exposing fresh surface to react with.

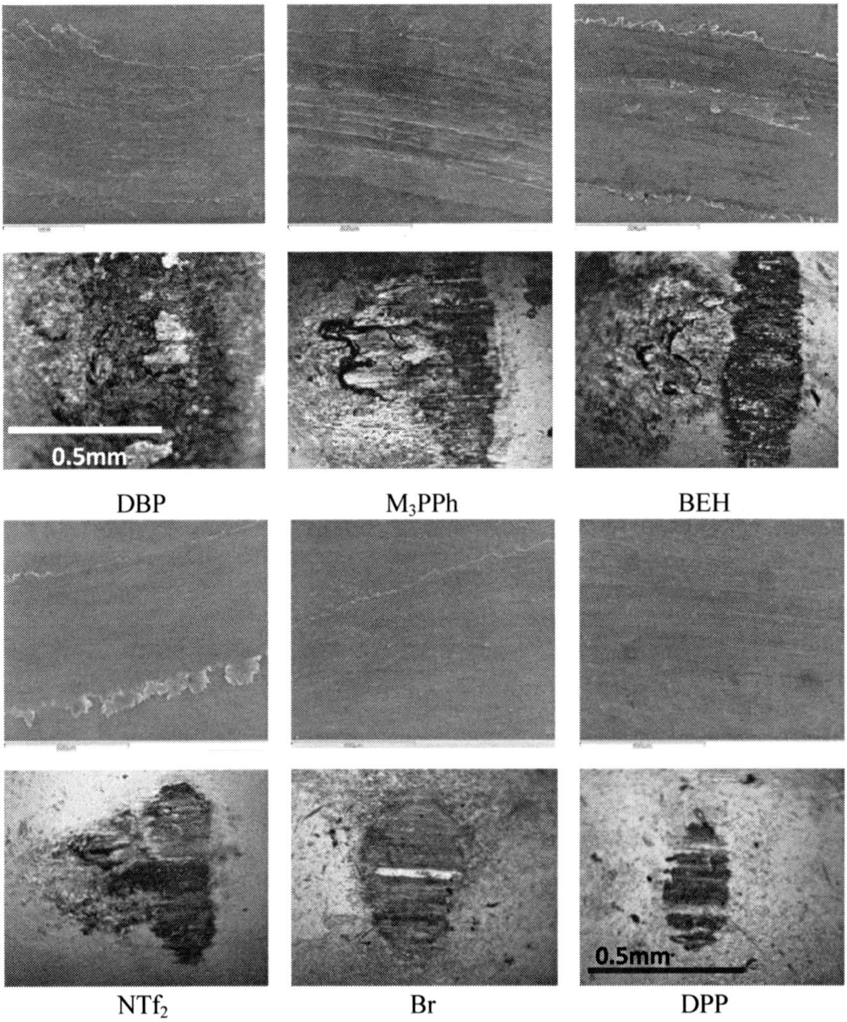

Figure 2: Wear scars on the aluminium disks and steel balls for the IL lubricated samples tested at 30N for 2500m.

The spectrum for $P_{66614}NTf_2$ shows the presence of fluorine and sulphur, both components of the IL anion. This suggests that it is the anion that has adsorbed onto the surface of the aluminium. The spectrum for $P_{66614}NTf_2$ also shows a very high oxygen content, suggesting that the film formed is part of an oxidation reaction. The IL spectrum for the lubricant that showed the lowest wear-coefficient, $P_{66614}DPP$, shows the presence of phosphorus on the aluminium surface. This spectrum is similar to that of $P_{66614}M_3PPh$ and $P_{66614}BEH$, but the IL lubricant resulted in a much lower wear coefficient. In this case the IL may not react with the surface to such an extent that tribocorrosion occurs. Bermudez

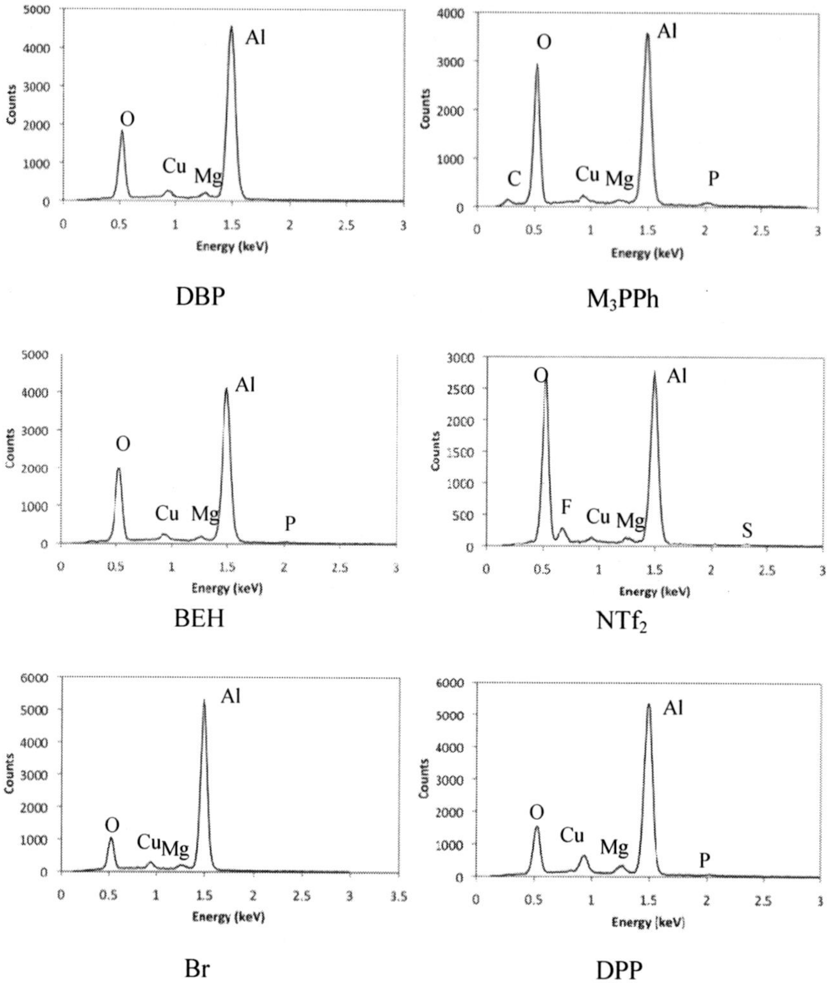

Figure 3: EDS spectra of wear scars on IL lubricated aluminium disks tested at 30N.

found that similar ILs had markedly different rates of wear and tribocorrosion, depending on their polarity.

In comparing the performance of the four phosphorus-containing IL lubricants there may be a correlation to their structure (Scheme 1) and viscosity (Table 1). In order from highest to lowest wear coefficient the novel IL lubricants are: P_{66614}DBP, $P_{66614}M_3$PPh, P_{66614}BEH and P_{66614}DPP. The DBP anion contains simple alkyl chains and has a viscosity of 130 mPa.s. The M_3PPh anion has short, branched chains and a similar viscosity, 120 mPa.s. The BEH anion has long branched chains, and its viscosity is much higher (260 mPa.s). The DPP anion has two phenyl rings present and a similar viscosity to the BEH,

210 mPa.s, but a much lower wear coefficient. It has been suggested that the ability of ILs to adsorb onto the surface and form layers is important for lubrication under film-forming conditions [4, 9] and the phenyl rings present in the DPP may influence its film forming ability, since it has achieved such a low wear coefficient. The DPP may be able to form layers on the surface due to surface interactions and/or the ability of the rigid phenyl rings to remain flat. The polarity of the ILs may also play a role and the complex nature of the films formed during wear needs further investigation [9].

4 Conclusion

The wear performance of four novel phosphonium based ionic liquids with phosphorus containing anions was assessed for lubricating ISO100Cr6 steel sliding on AA2024 aluminium disks. Their performance was evaluated in comparison to fully formulated diesel engine oil and similar phosphonium ionic liquids containing bromide and NTf_2anions. The performance of these novel ILs varied widely, with the P_{66614}DPP lubricated sample achieving the lowest wear coefficient and the P_{66614}DBP the highest. It is suggested that the phenyl rings in the structure of P_{66614}DPP may be beneficial in forming a layer structure on the contact surfaces that reduces friction and wear.

References

[1] Jiménez, A.-E. &Bermúdez, M.-D., Ionic liquids as lubricants for steel–aluminum contacts at low and elevated temperatures.*Tribology Letters*, 2007. **26(1)**: pp. 53-60.

[2] Mu, Z., Wang, X., Zhang, S., Liang, Y., Bao, M., &Liu, W., Investigation of tribological behavior of Al--Si alloy against steel lubricated with ionic liquids of 1-diethylphosphonyl-n-propyl-3- alkylimidazolium tetrafluoroborate. *Journal of Tribology*, 2008. **130(3)**: pp. art. no. 034501.

[3] Qu, J., Blau, P.J., Dai, S., Luo, H., Meyer III, H.M., &Truhan, J.J., Tribological characteristics of aluminum alloys sliding against steel lubricated by ammonium and imidazolium ionic liquids.*Wear*, 2009. **267(5-8)**: pp. 1226-1231.

[4] Liu, X., Zhou, F., Liang, Y., &Liu, W., Tribological performance of phosphonium based ionic liquids for an aluminum-on-steel system and opinions on lubrication mechanism.*Wear*, 2006. **261(10)**: pp. 1174-1179.

[5] Kenbeek, D. &Buenemann, T.F., *Lubricant additives: chemistry and applications*. 1st ed, ed. L.R. Rudnick. 2003, Boca Raton, FL: CRC Press.

[6] Mu, Z., Zhou, F., Zhang, S., Liang, Y., &Liu, W., Effect of the functional groups in ionic liquid molecules on the friction and wear behavior of aluminum alloy in lubricated aluminum-on-steel contact.*Tribology International*, 2005. **38(8)**: pp. 725-731.

[7] Jiménez, A.E., Bermúdez, M.D., Iglesias, P., Carrión, F.J., &Martínez-Nicolás, G., 1-N-alkyl -3-methylimidazolium ionic liquids as neat

lubricants and lubricant additives in steel-aluminium contacts. *Wear*, 2006. **260(7-8)**: pp. 766-782.

[8] Nicholls, M.A., Do, T., Norton, P.R., Kasrai, M., &Bancroft, G.M., Review of the lubrication of metallic surfaces by zinc dialkyl-dithiophosphates. *Tribology International*, 2005. **38(1)**: pp. 15-39.

[9] Bermúdez, M.D. &Jiménez, A.E., Surface interactions and tribochemical processes in Ionic Liquid lubrication of aluminium-steel contacts. *International Journal of Surface Science and Engineering*, 2007. **1(1)**: pp. 100-10.

[10] Hu, L., Chen, J., Liu, W., Xue, Q., &Kajdas, C., Investigation of tribochemical behavior of Al-Si alloy against itself lubricated by amines. *Wear*, 2000. **243(1-2)**: pp. 60-67.

[11] Howlett, P.C., Zhang, S., MacFarlane, D.R., &Forsyth, M., An investigation of a phosphinate-based ionic liquid for corrosion protection of magnesium alloy AZ31. *Australian Journal of Chemistry*, 2007. **60(1)**: pp. 43-46.

[12] Zhang, Q.B. &Hua, Y.X., Corrosion inhibition of mild steel by alkylimidazolium ionic liquids in hydrochloric acid. *Electrochimica Acta*, 2009. **54(6)**: pp. 1881-1887.

[13] Jiménez, A.E., Bermúdez, M.D., Carrión, F.J., &Martínez-Nicolás, G., Room temperature ionic liquids as lubricant additives in steel-aluminium contacts: Influence of sliding velocity, normal load and temperature. *Wear*, 2006. **261(3-4)**: pp. 347-359.

[14] Qu, J., Truhan, J., Dai, S., Luo, H., &Blau, P., Ionic liquids with ammonium cations as lubricants or additives. *Tribology Letters*, 2006. **22(3)**: pp. 207-214.

[15] Yagi, T., Sasaki, S., Mano, H., Miyake, K., Nakano, M., &Ishida, T., Lubricity and chemical reactivity of ionic liquid used for sliding metals under high-vacuum conditions. *Proceedings of the Institution of Mechanical Engineers, Part J: Journal of Engineering Tribology*, 2009. **223(7)**: pp. 1083-1090.

[16] Ye, C., Liu, W., Chen, Y., &Yu, L., Room-temperature ionic liquids: a novel versatile lubricant. *Chemical Communications*, 2001**(21)**: pp. 2244-2245.

[17] Bermúdez, M.D., Jiménez, A.E., Sanes, J., &Carrión, F.J., Ionic liquids as advanced lubricant fluids. *Molecules*, 2009. **14(8)**: pp. 2888-2908.

[18] Chen, Yunxia, Ye, Chengfeng, Wang, Haizhong, Liu, and Weimin, Tribological performance of an ionic liquid as a lubricant for steel/aluminium contacts. *Journal of Synthetic Lubrication*, 2003. **20(3)**: pp. 217-225.

[19] Mistry, K., Fox, M.F., & Priest, M., Lubrication of an electroplated nickel matrix silicon carbide coated eutectic aluminium–silicon alloy automotive cylinder bore with an ionic liquid as a lubricant additive. *Proceedings of the Institution of Mechanical Engineers, Part J: Journal of Engineering Tribology*, 2009. **223(3)**: pp. 563-569.

[20] Sun, J., Howlett, P.C., MacFarlane, D.R., Lin, J., & Forsyth, M., Synthesis and physical property characterisation of phosphonium ionic liquids based on $P(O)_2(OR)_2^-$ and $P(O)_2(R)_2^-$ anions with potential application for corrosion mitigation of magnesium alloys. *Electrochimica Acta*, 2008. **54(2)**: pp. 254-260.
[21] McNulty, J., Cheekoori, S., Bender, T.P., & Coggan, J.A., A pronounced anionic effect in the Pd-catalyzed Buchwald-Hartwig animation reaction revealed in phosphonium salt ionic liquids. *European Journal of Organic Chemistry*, 2007(**9**): pp. 1423-1428.
[22] Bradaric, C.J., Downard, A., Kennedy, C., Robertson, A.J., & Zhou, Y., Industrial preparation of phosphonium ionic liquids. *Green Chemistry*, 2003. **5(2)**: pp. 143-152.

Author Index